Analysis of Environmental Endocrine Disruptors

ACS SYMPOSIUM SERIES 747

Analysis of Environmental Endocrine Disruptors

Lawrence H. Keith, EDITOR
Waste Policy Institute

Tammy L. Jones-Lepp, EDITOR
United States Environmental Protection Agency

Larry L. Needham, EDITOR
Centers for Disease Control and Prevention

American Chemical Society, Washington, DC

Library of Congress Cataloging-in-Publication Data

Analysis of environmental endocrine disruptors / Lawrence H. Keith, Tammy L. Jones-Lepp, Larry L. Needham, editors.

p. cm.—(ACS symposium series , ISSN 0097–6156 ; 747)

Includes bibliographical references and index.

ISBN 0–8412–3650–X

1. Environmental toxicology Congresses. 2. Endocrine toxicology Congresses.

I. Keith, Lawrence H., 1938– . II. Jones-Lepp, Tammy L. III. Needham, Larry L. IV. Series.

RA1226.A53 2000
615.9′02—dc21

99–42831
CIP

The paper used in this publication meets the minimum requirements of American National Standard for Information Sciences—Permanence of Paper for Printed Library Materials, ANSI Z39.48–1984.

Distributed by Oxford University Press

PRINTED IN THE UNITED STATES OF AMERICA

Advisory Board

ACS Symposium Series

Foreword

THE ACS SYMPOSIUM SERIES was first published in 1974 to provide a mechanism for publishing symposia quickly in book form. The purpose of the series is to publish timely, comprehensive books developed from ACS sponsored symposia based on current scientific research. Occasionally, books are developed from symposia sponsored by other organizations when the topic is of keen interest to the chemistry audience.

Before agreeing to publish a book, the proposed table of contents is reviewed for appropriate and comprehensive coverage and for interest to the audience. Some papers may be excluded in order to better focus the book; others may be added to provide comprehensiveness. When appropriate, overview or introductory chapters are added. Drafts of chapters are peer-reviewed prior to final acceptance or rejection, and manuscripts are prepared in camera-ready format.

As a rule, only original research papers and original review papers are included in the volumes. Verbatim reproductions of previously published papers are not accepted.

ACS BOOKS DEPARTMENT

Contents

Preface

With the public's first collective recognition of the direct linkages between the industrial world and the natural environment, catalyzed largely by the publication of *Silent Spring* in 1962, new environmental concerns cycle through public and scientific debate at a seemingly escalating pace. An international dialog was begun in 1962 with concern over the use of DDT and its adverse effects on predatory birds; Times Beach and dioxin then became household words—followed by the apples-and-Alar scare.

A third of a century later, with the 1996 publication of *Our Stolen Future,* the most recent focus of concern is on chemicals that might have adverse effects on endocrine systems—in both humans and other organisms. This chemically diverse suite of substances has been named endocrine disrupting chemicals (EDCs). As with prior debates among believers and skeptics regarding the significance of chemicals in the environment, eventually yielding to concerted scientific research, EDCs are experiencing the same type of discourse. At one extreme, the believers maintain that EDCs portend the end of human reproduction. At the other, the skeptics state that naturally occurring bioactive chemicals in our diet are of more concern than anthropogenic chemicals.

Many questions surround EDCs. Are these chemicals at trace environmental concentrations adversely affecting endocrine systems? Are human sperm counts down? Is the female-to-male birth ratio becoming skewed? Do these chemicals act alone or does a synergistic effect occur? How widespread is the occurrence of these chemicals, and what are their ultimate environmental fates? What chemicals are actually EDCs, and which are not? New questions continue to emerge at a pace exceeding science's ability to address them.

The 216th American Chemical Society National Meeting presented an opportunity to convene a special ACS symposium, in Boston, that highlighted the issues surrounding EDCs. This book assembles from the symposium a collection of chapters addressing effects and emerging issues involving EDCs on humans, wildlife, and research animals, the results of current efforts to develop screening methods for EDCs, current efforts to develop confirmatory monitoring and measurement methods to complement the results from the screening methods, and current updates on regulatory and policy issues involving monitoring and regulation of EDCs in the environment and food. We thank the authors for contributing their well-written and thought-provoking chapters for this book. It is our hope that through broadened and vigorous dialog some of the mystery surrounding EDCs can be brought out into the scientific light and made less mysterious to both the scientific and lay community. We will need to approach the issues surrounding EDCs in a holistic manner by examining the intricacies of the chemical–exposure–effect continuum.

LAWRENCE H. KEITH
Waste Policy Institute
Suite 2100, 2000 Draft Drive
Blacksburg, VA 24060–6354

TAMMY L. JONES-LEPP
U.S. Environmental Protection Agency
P.O. Box 93478
Las Vegas, NV 89193–3478

LARRY L. NEEDHAM
Centers for Disease Control and Prevention
4770 Buford Highway, Mailstop F17
Atlanta, GA 30341

Chapter 1

Endocrine Disruptor Screening and Testing: A Consensus Strategy

G. E. Timm and A. F. Maciorowski

Office of Prevention, Pesticides, and Toxic Substances, U.S. Environmental Protection Agency, Washington, DC 20460

Over the past five years, endocrine disrupting chemicals have emerged as a major environmental science and policy issue. Documented effects to fish and wildlife populations, coupled with evidence form human poisonings, epidemiology, and experimental toxicology have led to an emerging hypothesis that chemicals may be affecting reproduction and development. In response to these concerns, the U.S. Congress passed the Food Quality Protection Act (FQPA) and the Safe Drinking Water Act (SDWA) Amendments of 1996. The FQPA requires EPA to screen pesticides for estrogenic effects on human health and permits EPA to screen chemicals found in drinking water sources for other hormonal effects. Faced with implementing a new regulatory program, EPA responded by sponsoring focused research to better understand the basic science of endocrine disruption, and by establishing the Endocrine Disruptor Screening and Testing Advisory Committee (EDSTAC) to advise the Agency on the design of an endocrine disruptor screening program.

Background

Scientific evidence has accumulated that humans, domestic animals, and fish and wildlife species have displayed adverse impacts form exposure to chemicals that interact with the endocrine system *(1, 2)*. To date, such problems have been best documented in humans, and domestic and wildlife species that have been exposed to high concentrations of organochlorine compounds (e.g., DDT and its metabolite DDE, PCBs, and dioxins), and to a few naturally occurring plant estrogens (*3*). However, whether such effects in the human population at-large can be attributed to environmental concentrations of contaminants remains unclear. Conflicting reports regarding declines in the quality and quantity of sperm production in humans have been published (*1*). Although reported increases in cancers of endocrine sensitive tissues (e.g., breast, prostate, testicular) are clear, direct evidence linking disease trends with exposure to environmental pollutants is scant. An endocrine-related basis for such effects is plausible, but considerable scientific uncertainty exists. Nevertheless, there is little doubt that small disturbances in endocrine function, particularly during certain highly sensitive stages of the life cycle (e.g., development, pregnancy, lactation), can lead to profound irreversible adverse effects (*3, 4*).

Taken collectively, the body of scientific research on human epidemiology, laboratory animals, fish and wildlife provides a plausible scientific hypothesis that environmental contaminants can disrupt the endocrine system leading to adverse health consequences. A critical issue is whether ambient environmental concentrations are sufficiently high to exert adverse effects on the general population. Epidemiology, mammalian toxicology, and ecological toxicology investigations are all necessary to resolve the fundamental scientific questions and uncertainties surrounding the endocrine disruptor issue.

The U.S. has implemented a two-part approach to endocrine disruptors. The first, is a research program to increase understanding of the basic scientific issues concerning endocrine disruption. The second is a regulatory screening and testing program to identify and characterize the effects of endocrine disrupting chemicals.

The Need for Research

A research agenda to assist the Federal government in making informed decisions regarding endocrine disruptors was developed for health and ecological effects at two workshops held in 1995 (*3, 5*). The following major research questions were identified from the workshops.

- What types of adverse effects are caused by endocrine disruptors?
- What chemicals are responsible for causing endocrine disruption?
- What exposure levels are required to cause adverse effects?
- Are effects on fish and wildlife due to isolated high exposure incidents or the result of a broader environmental contamination problem?
- How much exposure do humans, fish and wildlife have to endocrine disrupting chemicals?
- What is the shape of the dose-response curve?

- Are breast and testicular cancers due to endocrine disruptors?
- Are test protocols adequate to detect the effects of endocrine disruptors?

Federal Research Inventory. The U.S. Office of Science and Technology Policy created a research inventory to coordinate research among 14 Federal agencies and identify gaps in ongoing research that should be addressed in a Federal research strategy (www.epa.gov/endocrine). The inventory lists 396 research projects of which nearly 70% are directed toward human health, with the remaining projects evenly split between ecological effects and exposure. One limitation of ongoing research evident from the inventory is the focus on relatively few chemicals such as PCB's, dioxins, and persistent pesticides. In addition, most of the projects deal with reproductive effects followed in decreasing order, by cancer, neurotoxicity and immunotoxicity. The majority of projects are basic research oriented, followed by hazard characterization, and the development of risk models, biomarkers, and exposure measurement methods. The Workgroup on Endocrine Disruptors identified three major research need areas: methods development; model development; and laboratory and field data measurements(*6*).

Research Priorities. The highest priority topics identified in the methods development category were to:

- Establish the effects of endocrine disruptors during the stages of development;
- Develop non-invasive biomarkers for wildlife;
- Develop techniques for assessing exposure during critical life stages; and
- Develop biomarkers of exposure for use in screening techniques.

The highest priority topics in the models development category were to:

- Develop methods to use mechanistic and biomarker data in risk models;
- Characterize dose-response relationships at environmentally realistic conditions;
- Develop risk models for ecological risk assessments; and
- Develop fate and transport models soil, air and water.

In the laboratory and field measurements category, the highest priority topics were to:

- Establish relationships between exposure and adverse endocrine effects;
- Develop exposure and effects data for amphibians, reptiles, and invertebrates; and

- Quantify endocrine disruptor body burdens in humans and wildlife.

The U.S. research inventory served as a prototype for an international research inventory developed by the World Health Organization and the International Program on Chemical Safety.

Requirement to Develop a Screening Program

Faced with increasing concern over endocrine disrupting chemicals and adverse human health outcomes, the U.S. Congress passed two laws that contained provisions for screening chemicals for their potential to disrupt the endocrine system. The Food Quality Protection Act of 1996 (P. L 104-170), 21 U.S.C. § 346a(p)) requires EPA to:

> "develop a screening program using, appropriate validated test systems and other scientifically relevant information, to determine whether certain substances may have an effect in humans that is similar to an effect produced by a naturally occurring estrogen, or such other endocrine effect as the Administrator may designate."

When carrying out the screening program, EPA (A) "shall provide for the testing of all pesticide chemicals" and (B) "may provide for the testing of any other substance that may have an effect that is cumulative to an effect of a pesticide chemical if the Administrator determines that a substantial population may be exposed to such a substance." 21 U.S.C. § 346a(p)(3). In addition, Congress amended the Safe Drinking Water Act (42 U.S.C. § 300j-17) authorizing the EPA to screen contaminants in drinking water to which substantial numbers of people would be exposed.

The congressional mandate to develop a screening program in a controversial and rapidly emerging area of science led EPA to establish The Endocrine Disruptor Screening and Testing Advisory Committee (EDSTAC) in October 1996. The EDSTAC was charged with advising the Agency on the development of a practical, scientifically defensible endocrine disruptor screening strategy. The EDSTAC consisted of 39 members representing the pesticide and chemical manufacturers, state and Federal government, and environmental and public health advocacy organizations.

At its first meeting, EDSTAC expanded the scope of its deliberations to include potential effects on the androgen and thyroid systems in addition to estrogen. The Committee cited numerous examples of anti-androgen and anti-thyroid agents and the impact that these systems have on reproduction, growth and development as reasons for their inclusion . Ecological effects were also deemed important in that ecological effects have provided the strongest evidence of endocrine disruption to date. Finally, EDSTAC also included chemicals other than pesticides and drinking water contaminants as candidates for screening and testing. The universe of candidate chemicals under consideration included approximately 900 pesticide active ingredients; 2,500 pesticide formulation inert ingredients; 75,500 industrial chemicals; and 8,000 cosmetics, food additives and nutritional supplements (7).

Screening for Endocrine Disruptors

The overall frame work that EDSTAC recommended consists of four major steps: (1) initial sorting of chemicals, (2) establishment of screening priorities, (3) Tier 1 screening, and (4) Tier 2 testing. The initial sorting step was intended to separate the chemicals into four discrete categories. Category 1 consists of polymers with a numerical average molecular weight (NAMW) greater than 1,000 daltons and certain exempted chemicals that are unlikely to display endocrine activity (e.g., certain biologically inactive ingredients or highly reactive substances such as strong mineral acids and bases that will react at the portal of entry). Polymers with the specified NAMW will be set aside pending a review of their monomers. Such polymers were thought to pose little risk because, they are generally too large to pass through biological membranes and interact with the endocrine system. Category 2 consists of chemicals with insufficient data to determine their potential for endocrine activity. Category 3 includes those chemicals that have sufficient data to bypass screening, but need testing. Finally, Category 4 consists of substances with adequate data which will be referred to the appropriate agency for hazard assessment.

Priority Setting. The largest category of chemicals, and that of greatest interest to EPA concerning screening, are those chemicals with insufficient information and data (Category 2). EDSTAC considered several different approaches for setting priorities among these chemicals. One consisted of ranking chemicals with known exposure and effects as the highest priority for screening and testing. EDSTAC was concerned, however, that such an approach would focus attention on the most studied chemicals, resulting in a low priority for chemicals that were little studied. The Committee also recognized that priority setting is not generally an objective, data-driven process. Comparable, equitable, and uniform data are simply not available for the majority of commodity chemicals. As an example, ranking a chemical that has extensive monitoring data in wildlife against one known to have human exposure, but entirely wanting for effects data in mammals, is not a simple task. The Committee wanted a system where like information and data could be compared in priority setting, but one that was also subject to clearly stated value judgments to ensure public confidence. The approach recommended by EDSTAC was termed a "compartment-based" priority setting strategy.

In the compartment-based strategy, a number of compartments (or sets) of chemicals are defined and the individual chemicals within each set are prioritized. One might think of mathematics and sets of numbers as an analogy. In mathematics one could define sets of real numbers, integers, irrational numbers, even numbers, etc. Numbers within each set could then be ordered in some fashion. Some numbers obviously belong to more than one set, and the same will hold true for chemicals. One could define a set of high production volume chemicals, chemicals measured in biota, chemicals in consumer products, chemicals detected in the workplace, etc., and then prioritize the chemicals within each group. Given the large number of chemicals that must be prioritized for screening, implementation will undoubtedly occur in phases or batches. Once the sets of chemicals are defined and prioritized, a batch of chemicals will be selected for the initial phase of the screening program. The contribution of each set of chemicals to this batch is the key subjective judgment that must be made in the priority setting process. The size of the first batch and spacing of subsequent batches of chemicals depends largely on the available laboratory capacity of the system, the ability of industry to pay for testing, and the resources of the EPA to review submitted data.

EDSTAC conceptualized the compartment-based priority setting approach, but did not determine what the sets should be. Rather, it advised EPA to further develop the concept and details of the approach with continued public involvement. EPA will implement the EDSTAC's recommendation by holding a workshop to define these sets and by completing the priority setting approach recommended by EDSTAC.

EDSTAC also recognized other problems for priority setting. Few chemicals actually have data that directly measures their endocrine disrupting potential. Most chemicals have scant effects data. Often these data are limited to short-term mutagenicity tests and acute toxicity in rodents. Even chemicals that have been examined with respect to reproduction and developmental effects were likely tested using conventional protocols that have not been specifically designed to detect endocrine effects. EDSTAC believed that the most expedient means for obtaining useful endocrine effects data was to employ high throughput screening technology to high production volume chemicals and pesticides. The first two *in vitro* assays recommended by EDSTAC may, in fact, be conducted in a high throughput mode. These assays are specific to receptor binding modes of action. However, all 15,000 high production volume chemicals could be assayed in approximately 3 to 6 months at a relatively modest cost. This information could be used with production and exposure information and data to assist priority setting for further screening. A positive result in the high throughput screen would raise the priority for testing a substance relative to its priority based purely on production and exposure. Concomitantly, a negative in the high throughput screen would neither raise nor lower the priority since a chemical could be endocrine active through mechanisms other than receptor binding.

Tier 1 Screening. EDSTAC built upon the work of other expert workgroups in reviewing candidate assays for a screening system (*8, 9*). These expert groups reviewed a number of individual assays and concluded that a battery of assays was necessary to evaluate endocrine disruption potential. They further recommended that the battery include *in vitro* and *in vivo* assays. *In vitro* assays are advantageous in that they are inexpensive and specific for a particular mode of action. Conversely, they lack the metabolic and response complexity of intact animals. EDSTAC recommended a battery of three *in vitro* assays and five *in vivo* assays for Tier 1 screening. The *in vitro* assays include an estrogen receptor binding or reporter gene assay, an androgen receptor binding or reporter gene assay, and a steroidogenesis assay using minced testes.

EDSTAC preferred functional assays over receptor binding assays because the former can be used to differentiate agonist from antagonist activity, and they are also more sensitive than the latter (*7*). Permanently transfected reporter gene assays, that is, assays that use cells permanently modified by the introduction of DNA and produce a particular marker protein, were recommeded by EDSTAC. Luciferase, the enzyme involved in the production of the flash in fireflies, was the specific marker chosen by EPA for the estrogen (ER) and androgen (AR) receptor assays it is developing. Thus, a positive response in the assay produces a flash of light that can be read photometrically. The reporter gene assays are capable of high throughput and will be run in the high throughput mode by EPA and used to prioritize chemicals for the rest of the assays.

The *in vivo* Tier 1 assays are:

Rodent 3-day Uterotrophic Assay
Rodent 20-day Pubertal Female Assay with Thyroid
Rodent 5-7 day Hershberger Assay
Frog Metamorphosis Assay
Fish Gonadal Recrudescense Assay

Deliberate endpoint complementarity is incorporated into the screening battery, and the assays in the battery are meant to work together as a whole. Thus, the ER reporter gene, uterotrophic and pubertal female assays screen for estrogenicity and anti-estrogenicity. The AR reporter gene and Hershberger assays screen for androgenicity and anti-androgenicity. The frog assay and pubertal female assays screen for thyroid. The fish assay is mainly present because fish are the class of vertebrates most distant from the mammals in terms of their metabolism and hormone systems. Having diverse taxa in Tier 1 may provide some idea of when we can be confident of consistent results among organisms and when we must be concerned about variability. A weight of the evidence approach will be used to evaluate the results in Tier 1. *In vivo* results will outweigh *in vitro* results. Chemicals testing negative in Tier 1 would be regarded as having low potential for interaction with the estrogen, androgen or thyroid systems. Chemicals testing positive would proceed for more in depth evaluation in Tier 2.

Tier 2 Testing. The Tier 2 tests are meant to identify adverse effects due to endocrine disruption and to establish a relationship between dose and response. The criteria identified by EDSTAC for Tier 2 tests was that the most sensitive life stage be tested (in utero or in ovo), that the tests be multi-generational and that each major taxonomic group be represented. Thus EDSTAC recommended five multi-generation tests: one each in mammals, birds, fish, amphibians and invertebrates. Tier 2 can be tailored based on both exposure and effects information. For instance, if it can be shown that only exposure to humans will occur, only the 2-generation test in rodents will be required.

EPA Implementation

EPA has begun to implement EDSTAC's recommendations. A pilot demonstration of the high throughput pre-screen was initiated in March 1998 to determine the suitability of the assay system for commercial chemicals and pesticides. Data should be available by early 1999. If the demonstration is successful, the full high throughput pre-screen could be started on 15,000 chemicals by the end of 1999, and priorities for further screening could be set in the year 2000. Meanwhile, one of EDSTAC's recommendations is that proposed screens and tests be standardized and validated to ensure that they give reliable and repeatable results. A taskforce has been established in the U.S. to coordinate the standardization and validation effort among government, industry and public interest groups. A workgroup under the Organization for Economic Cooperation and Development (OECD) will coordinate international efforts to standardize and validate screens and tests that are of interest to a number of member countries. It is estimated that two years will be required to standardize and validate the screens and perhaps five years will be required for some of the newer tests.

EPA published a notice in the Federal Register on December 28, 1998, to obtain comments on its proposed screening program and implementation plans. The proposed screening program and the implementation plan will also be reviewed by the EPA Science Advisory Board and FIFRA Scientific Advisory Panel in March 1999.

Low Dose Effects

One issue raised, but not resolved by EDSTAC, was the concern over low dose effects. Specifically, can *in utero* exposure to environmental levels of endocrine disrupting chemicals result in permanent developmental changes? If so, what is the shape of the dose-response curve? Does it have a threshold? Can the effects at low doses be inferred from experimental test results at high doses?

This is a highly controversial area and a high priority for research. Some scientists believe that the answer to the first question is a clear yes. They also argue that one cannot predict the effects that might occur at low doses based upon high dose testing because different mechanisms or modes of action may be involved. If true, this would have profound implications for testing, risk assessment and the management of chemical risks. Test guidelines would need to be changed to include low doses. If testing at low doses is necessary, should the guidelines continue to require that the highest dose be set at "maximally tolerated" levels? If so, we would need to add several additional dose levels to the three that are already required by EPA and OECD test guidelines. Whether the endpoints being measured at the low doses are sufficiently sensitive that the numbers of animals required in current guidelines are adequate, or whether the numbers of animals have to be to increased is another question. Such issues are difficult from a test design standpoint.

From a risk assessment and risk management perspective, low dose endocrine effects might be dealt with in much the same way we have dealt with cancer. We may conclude that there is some risk at any exposure. In some cases this would mandate such low control levels that chemicals would be forced off the market. The U.S. government, industry and environmental groups are in the process of designing and implementing a research program to investigate low dose effects.

Conclusions

Many policy issues in the United States and abroad will be decided before the resolution of all scientific questions. Nevertheless, we see that the results of screening and testing and the answers to many of the questions posed by the research agenda will combine to form our conclusions regarding the level of risk that endocrine disruptors pose to humans and wildlife. Ultimately in the United States, risk management actions will be taken under current U.S. laws – laws that are mainly risk based and, therefore, which require the integration and interpretation of effects and exposure data.

References

1. *European Workshop on the Impact of Endocrine Disruptors on Human Health and Wildlife*; EUR 17549; European Commission: Brussels, Belgium, 1996

2. *Expert Workshop on Endocrine Modulators and Wildlife: Assessment and Testing*, Tattersfield, L. et al. Eds.; SETAC-Europe; Brussels, Belgium, 1997.

3. Kavlock RJ et al. *Environmental Health Perspectives*. **1996**, 104, 715-740

4 *Special Report on Environmental Endocrine Disruption: An Effects Assessment and Analysis*. Environmental Protection Agency: Washington, DC, EPA/630/R-96/012, 1997.

5. Ankley, GT. et al. *Review of Toxicology*. **1997**, 1,71-106.

6. *A Framework for Planning*. National Science and Technology Coil, Committee on Environment and Natural Resources, Washington, DC, 1996.

7. *Endocrine Disruptor Screening and Testing Advisory Committee: Final Report;* Environmental Protecti Agency: Washington,DC , 1998. Available at www.epa.gov/opptintr/opptendo.

8. Ankley, G.T.; et al. *Environmental Toxicology and Chemistry*. **1998**, *17*, 68-87.

9. Gray LE et al. *Reproductive Toxicology.* **1997**, *11*, 719-750.

Chapter 2

Method Development and Interspecies Comparison of Estrogen Receptor Binding Assays for Estrogen Agonists and Antagonists

Nigel J. Bunce, Brian J. Cox, and Anthony W. Partridge

Department of Chemistry and Biochemistry, University of Guelph, Guelph, Ontario N1G 2W1, Canada

We have used gel filtration chromatography to separate specifically bound from nonspecifically bound and unbound radioligand in estrogen receptor binding assays. Relative binding affinities (RBAs) were obtained by the competition method, using [^{3}H]-estradiol as the radioligand. Mixtures of environmental estrogens behaved additively in the assay, and in an interspecies comparison, the order of affinity was similar in rat liver and trout liver. Similar RBAs in different species are supported by sequence alignment analysis, which indicates conservation of key amino acid residues and sequences in the ligand binding domains of estrogen receptors of diverse species. The results are discussed in the context of the requirement for screening assays for endocrine modulators.

Endocrine modulators (also called endocrine disrupting compounds, EDCs) can cause adverse effects by interfering with endocrine homeostasis. Direct acting EDCs interfere with some step in the mechanism of action of the normal hormone, for example by binding to the steroid hormone receptor or by altering subsequent downstream events in signal transduction. Indirect acting EDCs alter the rates of synthesis, secretion, transport, uptake, metabolism, or clearance of the steroid hormone. In the case of estrogenic activity, the subject of this contribution, endocrine modulation can cause both reproductive and developmental effects; these have been documented in all vertebrate orders, but are species-dependent.

Screening assays for estrogenic activity are needed for two quite separate applications: testing commercial chemicals, and screening environmental samples. In the case of environmental samples, mixtures of estrogenically active compounds, possibly including both agonists and antagonists, are almost always involved. Under the Food Quality Protection Act (1996) and 1996 amendments to the Safe Drinking Water Act, the United States Environmental Protection Agency (EPA) must develop and implement tests for detecting endocrine modulating effects of commercial chemicals. Besides pharmaceuticals and pesticides, which are covered under other legislation, this will require ~70,000 currently used chemicals and about 1500 new commercial introductions per year to be tested (1). Testing programs will require a combination of validated in vitro and in vivo tests, with clearly defined endpoints, appropriate to important stages of the life cycle (2,3). The large number of substances (and environmental samples) to be screened demands assays that are rapid, inexpensive, and capable of high throughput, in order to identify a smaller subset of samples that require more detailed testing. In vitro methods will be used of necessity for the preliminary screen, because they are faster, cheaper, and technically easier to perform than in vivo assays, even though they lack the biological plausibility of whole animal tests. Estrogen receptor (ER) binding assays are attractive candidates, because binding to the ER is the first step in the chain of events leading to direct estrogenic effects: i.e., they are mechanism based. Direct acting estrogen agonists and estrogen antagonists both give positive responses in ER binding assays, and so both can be considered "estrogenically active".

Receptor-ligand binding assays are employed in mechanistic biochemistry and toxicology to assess the relative binding affinities (RBAs) of ligands, with a view to defining parallels between RBA and biological potency. Binding affinity can be determined by means of a **saturation assay** or a **competition assay**. In a saturation assay, increasing concentrations of the test substance (ligand) are incubated with a fixed amount of ER, giving increasing amounts of bound ligand. The conventional version of the saturation assay is applicable only to radiolabelled ligands, but this limitation can be overcome by using reporter gene assays, in which binding to the ER induces a response from another gene under the control of the ER, such as light emission from firefly luciferase (2,4). Reporter gene assays are technically demanding, however, in terms of the need for stably transfected cell lines.

RBAs are commonly determined by means of the competition assay, in which a fixed concentration of radioligand and varying concentrations of an unlabelled competitor (or mixture) compete for the binding sites of a fixed concentration of ER. The greater the concentration of the unlabelled competitor, the more radioligand is displaced from the ER, and the less the bound radioactivity. The usual objective of a competition assay is to determine the EC_{50}, the concentration of competitor that displaces 50% of the radioligand. At the EC_{50}, the radioligand and the competitor have equal occupancies of the ER, and so their RBAs are in the ratio of their concentrations. Alternatively, a "positive/negative" answer can be obtained by selecting a cut-off RBA as the limit of a positive response. For example, suppose that estradiol is the reference radioligand and that 10^{-5} is the cut-off RBA. For a concentration of the reference radioligand of 10 nM (a reasonable value), a series of commercial chemicals to be tested could be prepared at 1 mM. The RBA will be below 10^{-5} if the bound radioactivity for radioligand + competitor is > 50% of the value for radioligand alone.

Hepatic ER was used in this work, because liver is an abundant source of ER and because liver contains only the ERα isoform, unlike uterus and ovary, which also contain the recently described ERβ (5,6). Selection of liver thus avoids the ambiguities associated with determining RBAs when two high-affinity receptors are present simultaneously.

Competitive receptor-ligand binding assays must be carried out under conditions that saturate the receptor's binding sites, so that the unlabelled competitor must compete with the radioligand for a deficiency of binding sites rather than simply bind to sites not already occupied by the radioligand. This requires the radioligand to be in excess over the receptor, and a method must be found to separate the "specifically bound" radioligand (those molecules bound to the ER binding sites) from "non-specifically bound" (low affinity binding to other proteins) and excess free radioligand. In commonly used assays such as the hydroxylapatite, protamine sulfate, and dextran coated charcoal (DCC) (7) assays (8), the total bound radioactivity (specific + non-specific) is separated from free radioligand by adsorption, but the non-specifically bound radioactivity must be determined in a separate experiment. In this work, we have developed a gel filtration chromatographic method to achieve baseline separation of the specific binding peak. Although chromatographic methods have previously been used to separate the ER for purification and diagnostic tests, they have not hitherto been applied to quantitative assays (9, 10, and references cited), possibly on account of lengthy separation times (> 1 h).

EXPERIMENTAL

Chemicals. Unlabelled chemicals were commercially available; [2,4,6,7-^{3}H]-17β-estradiol was purchased from Amersham in 250 μCi size with specific activities ranging from 85-105 Ci/mmol. The contents of the vial were quantitatively transferred in toluene to a glass 20 mL scintillation vial with a screw-top lid and blown down to dryness with dry N_2., then redissolved in DMSO to give a final concentration of 1 μM. The concentration was checked by measuring a 10 μL aliquot by scintillation counting.

Estrogen receptor. Cytosol containing ERα was prepared from the livers of female Long-Evans rats (95-105 g: Charles River) or female rainbow trout. The rats were sacrificed by CO_2 asphyxiation and cervical dislocation (trout were sacrificed by a blow to the skull and severing the spinal cord). The livers were perfused with ca. 10 mL of ice-cold buffer (25 mM Na_2HPO_4/NaH_2PO_4, 10 mM EDTA, 10 % glycerol and 100 mM KCl, adjusted to pH 7.1 with HCl) through the hepatic portal vein to remove blood. The weighed livers were resuspended in 2 mL of cold buffer per gram of liver, coarsely cut, and homogenized on ice. The homogenate was centrifuged at 9600 rpm (10,000 *g*) for 20 min, and the supernatant was then carefully removed and re-centrifuged for 100 min at 38,000 rpm (104,000 *g*). Surface lipid was then removed by suction, and the protein concentration of the cytosol was measured by the method of Bradford (11). The cytosol was then divided into 1 and 2 mL aliquots, frozen at –20 °C for 1 h, and stored at –70 °C.

HPLC Gel Filtration. The equipment comprised a Waters 6000A pump (eluting at 0.5 mL min^{-1}), a Rheodyne 7010 injector with a 100 μL injection loop, and Supelco GFC 1300 (300 x 7.5 mm, PEEK) or a Supelco Progel™ TSK G4000$_{SWXL}$ (300 x 7.8

mm, stainless steel) columns. All components were connected by PEEK (polyethylether ketone) 0.03 inch bore tubing and PEEK nuts and ferrules. Fractions of 0.5 mL were collected using a BioRad fraction collector model 2110 into Fisher Brand polypropylene 1.5 mL microcentrifuge tubes (with the snap cap cut off at the tube mouth), placed inside borosilicate test tubes (13 x 100 mm) in the fraction collector carrousel. The lip on the mouth of the microfuge tube rested on the mouth of the test tube, so that the microfuge tube with the effluent fraction could be transferred directly to the scintillation vial.

Data Processing. Values from the scintillation counting were manually entered into a spreadsheet and saved as ASCII files. These were manipulated to be compatible with the program Peak Simple 2 version 3.54 (SRI Instruments, Torrance, CA USA), which was used to integrate the peaks from the gel filtration chromatogram. The peak areas were entered into a Quattro Pro version 6.0 (Novell Inc.) spreadsheet for data analysis and statistics. The EC_{50} of the competitor was obtained from linear regression of "probit of percent binding" (100% was the value with no competitor present) *vs.* log concentration of the competitor, at the estimated value of "probit = 5.000". Relative Binding Affinity (RBA) was calculated as {(EC_{50} of competitor)/(EC_{50} of unlabelled 17β-estradiol)}(12).

Receptor meta-analysis. ER α sequences used were human (*Homo sapiens*, accession number 544257), Norway rat (*Rattus norviegicus*, 119600), chicken (*Gallus gallus*, 119597), African clawed frog (*Xenopus laevis*, 625330), Japanese eel (*Anguilla japonica*, 2073113), red sea bream (*Chrysophrys major*, 3122078), pig (*Sus scrofa*, 2500908), Medaka (*Oryzias sp.*, 3097501), Zebra finch (*Taeniopygia guttata*, 1449146), West African cichlid (*Oreochromis aureus*, 2507414), Atlantic salmon (*Salmo salar*, 1706708), rainbow trout (*Oncorhynchus mykiss*, 103903), Japanese medaka (*Oryzias latipes*, 1706707), mouse (*Mus musculus*, 119599). ER β sequence was from the Norway rat (*Rattus norviegicus*, 3077649). Sequences were found through PubMed at http://www.ncbi.nlm.nih.gov/PubMed/ using GenBank protein sequences. Multiple sequence alignment was done using PIMA by maximal linkage clustering and sequential branching clustering through the web site at the Baylor College of Medicine http://dot.imgen.bcm.tmc.edu:9331/cgi-bin/multi-align

Results and Discussion

Gel filtration Figure 1 shows a typical gel filtration chromatogram from the incubation of rat liver cytosol with [^{3}H]-estradiol. The peak eluting with 5.5 mL of buffer (M ≈ 1000 kDa) represents the high-affinity, low-capacity ERα, and could be completely out-competed with a 200-fold excess of unlabelled estradiol. The low-affinity high-capacity peak at V = 10 mL (M ≈ 50 kDa) was hardly competed, and did not separate completely from unbound estradiol (V ≈ 12 mL). Chromatograms obtained in experiments with rainbow trout liver cytosol had a similar appearance. Preliminary experiments indicated that a 30 min incubation at 30 °C was sufficient to maximize the area of the ERα peak. Saturation curves showed that the high-affinity, low-capacity rat hepatic ERα was fully saturated by 5 nM [^{3}H]-17β-estradiol at

8 mg/mL protein, but the low affinity protein was not close to saturation even at >10 nM [^{3}H]-17β-estradiol.

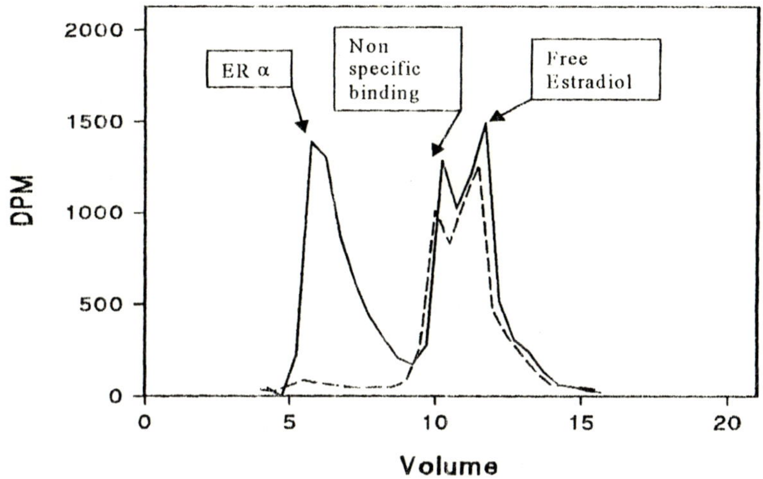

Figure 1: Gel filtration chromatogram of rat ERα incubated with 10 nM [^{3}H]-estradiol in the absence (solid line) and presence (dashed line) of 2 μM unlabelled estradiol. Reproduced with permission from Reference 13.

Competitions and EC_{50}. Figure 2 shows a sample probit curve for the competition between 10 nM [^{3}H]-estradiol and 4′-hydroxy-2,3,4,5-tetrachlorobiphenyl. Table 1 shows the EC_{50} values and regression parameters of the probit plots. The slopes were all between -1 and -1.5, except for dieldrin, a very weakly binding ligand. They were statistically indistinguishable by the Student's t-test (estradiol = −1.26, average slope = −1.29, 95% CI −1.39 to −1.13 at $^{\alpha}/_2 = 0.025$); this indicates that all these ligands competing for the same binding site on the ER (*cf.* Gaido and co-workers (14)). The ligands having two hydroxyl groups (estradiol, zearalenol, DES and estriol) yield slightly larger slopes than those lacking this structural feature (o,p'-DDT, estradiol-3-benzoate and estrone), probably due to differences in the points of contact made between the competitors and the ER binding site (15, 16).

The experimental uncertainty in EC_{50} values has been discussed by Schneider et al. (17), who studied competitive binding of dioxin-like compounds to the Ah receptor. They found that probit analysis of competitions with 5 to 6 concentration points spanning 4 to 5 orders of magnitude, as used by many authors, gave EC_{50}s whose 95 % CI covered more than two orders of magnitude. It required six runs of a fifteen point curve (total of 90 points) to reduce the 95 % CI to a factor of two on each side of the EC_{50}. These uncertainties make it not meaningful to dwell on small differences in EC_{50} values; we would not claim any detectable difference in EC_{50} (and therefore RBA) between, for example, α-zearalenol, estriol, and estrone.

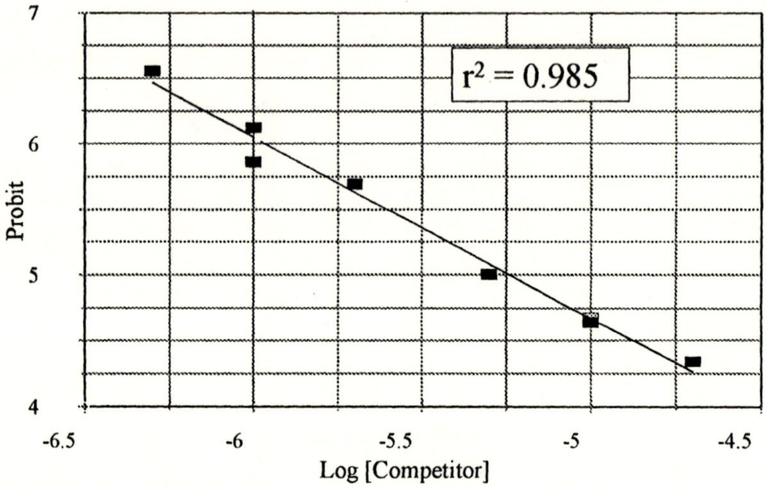

Figure 2: Probit plot for the competition between 10 nM [^{3}H]-estradiol and tetrachlorobiphenylol (only those data points having 4 < probit < 7 are included)

Table 1: EC_{50} of estrogenic compounds towards rat hepatic ERα

Compound	EC_{50}, nM	- Slope	Intercept	r^2
Ethynyl-estradiol	13	1.02 ± 0.094	-3.00 ± 0.251	0.951
17β-estradiol	25	1.26 ± 0.055	-4.56 ± 0.086	0.990
Diethylstilbestrol	28	1.54 ± 0.150	-6.64 ± 0.239	0.963
Estradiol-3-benzoate	60	1.11 ± 0.088	-2.98 ± 0.206	0.969
α-zearolenol	224	1.53 ± 0.085	-5.17 ± 0.149	0.988
Estrone	300	1.09 ± 0.036	-2.10 ± 0.097	0.994
Estriol	490	1.43 ± 0.072	-4.04 ± 0.102	0.988
Tetrachlorobiphenylol	5900	1.37 ± 0.069	-2.20 ± 0.105	0.985
o,p-DDT	2.2×10^5	1.01 ± 0.184	1.31 ± 0.283	0.937
Dieldrin	3.9×10^5	0.54 ± 0.105	3.15 ± 0.234	0.930

Comparing RBAs from different assays. Table 2 compares the RBAs from this work with literature values (17β-estradiol = 1.00). The overall order of RBAs is broadly similar, despite differences in experimental protocol. Kuiper *et al.* (6) expressed ERα *in vitro* from human ERα cDNA, and then used the purified protein in a competitive binding assay. Gaido et al. (14) and Coldham et al. (18) both used a yeast cell line expressing recombinant ERα with activation of a reporter gene.

Table 2: Comparison of relative binding affinity (RBA) values from this work with those from other techniques

Compound	RBA this work	RBA[6]	RBA[14]	RBA[18]
Estradiol	1.00	1.00	1.00	1.00
Diethylstilbestrol	0.89	4.68	0.64	0.74
α-zearolenol	0.42	-	-	0.087
Estrone	0.08	60	-	0.096
Estriol	0.05	14	3.7×10^{-3}	6.3×10^{-3}
Tetrachlorobiphenylol	4×10^{-3}	-	-	8.2×10^{-3}
o,p-DDT	1×10^{-4}	--	1×10^{-7}	1×10^{-6}

Additivity of mixtures. Table 3 shows the data from three mixtures, with the concentrations expressed in 17β-estradiol equivalents (molar ratios were calculated to give each component equal estradiol equivalent concentration *i.e.*, concentrations normalized to those of estradiol, based on relative EC_{50}s). The t-test showed that the EC_{50}s of the mixtures, expressed in estradiol equivalents, were indistinguishable, although the individual slopes and intercepts were distinct at $^{\alpha}/_{2} = 0.025$. The aggregated data provide good evidence that binding of these ligands to the ERα is an additive phenomenon (EC_{50} = 28 nM, $r^2 = 0.926$).

Interspecies comparison. In order for RBAs to be useful as a screening parameter for endocrine modulators, they should have applicability across a range of species,

Table 3: Data from competition of mixtures for the rat hepatic ERα

Compounds	EC_{50}	- Slope	-Intercept	r^2
estradiol (alone)	25 nM	1.26 ± 0.055	4.56 ± 0.086	0.990
estradiol (1) estradiol benzoate (2.4)	28	1.37 ± 0.228	5.33 ± 0.154	0.963
estradiol (1) α-zearolenol (8.9)	21	1.36 ± 0.189	5.44 ± 0.181	0.945
estradiol (1) tetrachlorobiphenylol (235)	39	1.40 ± 0.188	5.39 ± 0.253	0.949

otherwise a new series of tests would be required for every species that was to be protected. Fielden et al. (19) found only 58% conservation in the ligand binding domain of ERα from several different animals, from a sequence alignment analysis, and concluded that RBAs might therefore vary substantially among species. Recent work by Brzozowski et al. (15) and Ekena et al. (16) showed that three essential residues in the ligand-binding domain (Glu 353 and Arg 394 at the 3-OH and His 524 at the 17-OH of estradiol) make contact with estradiol and other ligands. Our own analysis of full receptor sequences of 13 ERα and one rat ERβ indicates that these residues are fully conserved in all fourteen receptors, along with a series of hydrophobic residues in the vicinity of the ligand binding site, while other substitutions in the neighbourhood are conservative (Figure 3).

Alignment #	390	391			435	436		556	570
Conserved	M LT AD		MI WAK ..	E LM GL W	S	PGKL F..	SNKGME		L K K
Mouse	MMGLLTNLAD	R	LVHMINWAKR..	LEILMIGLVW	SM	EHPGKLLF..	MSNKGME		LYNMKCK
Rat	MMGLLTNLAD	R	LVHMINWAKR..	LEILMIGLVW	SM	EHPGKLLF..	MSNKGME		LYNMKCK
Pig	MMGLLTNLAD	R	LVHMINWAKR..	LEILMIGLVW	SM	EHPGKLLF..	MSNKGME		LYNMKCK
Human	MMGLLTNLAD	R	LVHMINWAKR..	LEILMIGLVW	SM	EHPGKLLF..	MSNKGME		LYSMKCK
Chicken	MMTLLTNLAD	R	LVHMINWAKR..	LEILMIGLVW	SM	EHPGKLLF..	MSNKGME		LYNMKCK
Finch	MMTLLTNLAD	R	LVHMINWAKR..	LEILMIGLVW	SM	EHPGKLLF..	MSNKGME		LYNMKCK
Frog	MMTLLTNLAD	R	LVHMINWAKR..	LEILMVGLIW	SV	EHPGKLSF..	MSNKGME		LYSMKCK
Sea beam	MMTLLTSMAD	K	LVHMIAWAKK..	LEVLMIGLIW	SI	HCPGKLIF..	MSNKGME		LYSMKCK
Cichlid	IMTLLTSMAD	K	LVHMITWAKK..	LEVLMIGLIW	SI	QCPGKLIF..	MSNKGME		LYSMKCK
Medaka	MMTLLTSMAD	K	LVHMIAWAKK..	LEVLMIGLIW	SI	HCPGKLIF..	MSNKGME		LYSMKCK
Salmon	MMTLLTSMAD	K	LVHMIAWAKK..	LEVLMIGLIW	SI	HCPGKLIF..	MSNKGME		LYSIKCK
Trout	MMTLLTSMAD	K	LVHMIAWAKK..	LEVLMIGLIW	SI	HCPGKLIF..	MSNKGME		LYSIKCK
Eel	MMMSLTNLAD	K	LVLMISWAKK..	LEVLMLGLMW	SV	DHPGKLIF..	LSNKGME		LSNMKRK
Rat beta	MMMSLTKLAD	K	LVHMIGWAKK..	MEVLMVGLMW	SI	DHPGKLIF..	ISNKGME		LLSMKCK

Figure 3: Sequence alignment analysis of fourteen estrogen receptors in the vicinity of the ligand binding site. Reproduced with permission from Reference 13.

The sequence alignment analysis suggested that the ordering of RBAs towards ERα (*e.g.* as "strong", "medium", or "weak" ligands) might be rather well maintained across vertebrate species. This hypothesis is borne out by the data in Figure 4, in which we compared the RBAs of five test compounds towards both rat and rainbow trout liver ERα. A fish species was chosen in order to provide a severe test, because the amino acid sequences of fish ERα diverge more from mammalian ERα than those of birds or amphibians. RBAs for hydroxytamoxifen and ICI 182384 were within an order of magnitude of estradiol (RBA = 1.00) in both species ("strong activity"); estrone was about an order of magnitude less active ("medium activity") and tetrachlorobiphenylol (HPCB) bound another order of magnitude more weakly ("weak activity").

Receptor binding assays cannot distinguish between agonists and antagonists, but this is unnecessary, and probably undesirable, in a preliminary screen whose objective is to identify estrogenically active substances, regardless of whether they are agonists or antagonists. Because both agonists and antagonists bind to the ER, and because we have demonstrated that binding is additive, antagonists cannot mask the estrogenic activity of agonists. Of the compounds in Figure 4, estradiol is the natural potent agonist, estrone is an endogenous agonist of medium potency, tetrachlorobiphenylol (HPCB) is a weak agonist, and ICI 182382 is a strong pure antagonist. The behaviour of hydroxytamoxifen is species dependent: it is an agonist in the mouse, antagonist in the chicken and exhibits both agonist and antagonist properties in rats and humans (20).

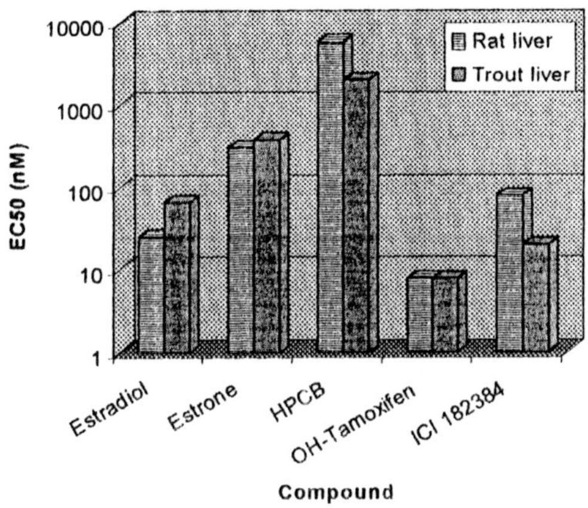

Figure 4: Comparison of RBAs of five different estrogenically active substances for rat and trout ERα

Conclusions

From the viewpoint of methodology, the advantage of the present assay is the use of gel filtration to separate the specific binding peak cleanly from binding to low affinity, high capacity proteins, and from free radioligand. Chromatographic methods of separation also allow the possibility of automated analysis, offering great savings in labour costs (the ER-ligand complexes are stable for several hours at 4 °C, and so numerous preparations could be loaded into an autosampler for overnight processing).

Concerning the specific application of the present work to xenoestrogens, the first requirement is to identify suspected endocrine modulators, which would then be subject to detailed testing (4, 19, 21). ER binding assays could be useful for screening commercial chemicals for estrogenic activity, because the ER is the first molecular target of both estrogenic and antiestrogenic substances, both of which would show positive in such a test. They could also be applied to environmental mixtures, and because mixtures of compounds behave additively towards the ERα, we can be confident that antagonists will not mask the activity of agonists.

An important conclusion is that RBAs appear to be well conserved across vertebrates, meaning that results obtained in one species should be applicable in others, at least to the level of assigning "strong", "medium", or "weak" estrogenic potency.

Acknowlegement We thank the Natural Sciences and Engineering Research Council of Canada for financial support.

References

1. US EPA (United States Environmental Protection Agency) (1997) Special Report on Environmental Endocrine Disruption: An Effects Assessment and Analysis, February 1997, Risk Assessment Forum, U.S. Environmental Protection Agency, Washington, D.C. 20460, Document Number EPA/630/R-96/012.
2. Ankley G.; Mihaich, E.; Stahl, R.; Tillitt, D.; Colborn, T.; McMaster, S.; Miller, R.; Bantle, J.; Campbell, P.; Denslow, N.; Dickerson, R.; Folmar, L.; Fry, M.; Giesy, J.; Gray, L E.; Guiney, P.; Hutchinson, T.; Kennedy, S.; Kramer, V.; LeBlanc, G.; Mayes, M.; Nimrod, A.; Patino, R.; Peterson, R.; Purdy, R.; Ringer, R.; Thomas, P.; Touart, L.; Van Der Kraak, G.; Zacharewski, T. (1998) Environ. Toxicol. Chem. 17, 68-87.
3. Gray, L.E. Jr.; Kelce, W.R.; Wiese, T.; Tyl, R.; Gaido, K.; Cook, J.; Klinefelter, G.; Desaulniers, D.; Wilson, E.; Zacharewski, T.; Waller, C.; Foster, P.; Laskey, J.; Reel, J.; Giesy, J.; Laws, S.; McLachlan, J.; Breslin, W.; Cooper, R.; Di Giulio, R.; Johnson, R.; Purdy, R.; Mihaich, E.; Safe, S.; Sonnenschein, C.; Welshons, W.; Miller, R.; MacMaster, S.; Colborn T. (1997) Reprod. Toxicol. 11, 719-750.
4. Zacharewski, T. (1997) Environ. Sci. Technol. 31, 613-623.
5. Mosselman, S.; Polman, J.; Dijkema, R. (1996) FEBS Letters 392, 49-53.
6. Kuiper, G.G.J.M.; Carlsson, B.; Grandien, K.; Enmark, E.; Haggblad, J.; Nilsson, S.; Gustafsson, J-A. (1997) Endocrinology 138, 863-870.
7. Shafie, S.; Brooks, S.C. (1979) J. Lab. Clin. Med. 94, 784-798.
8. Pavlik, E.J.; Coulson, P.B. (1975) J. Steroid Biochem. 7, 357-368.
9. Sluyser, M.; Wittliff, J.L. (1992) Mol. Cell. Endocrinol. 85, 83-88.
10. Folk, P.; Dong J.; Wittlife, J.L. (1992). J. Steroid Biochem. Mol. Biol. 42, 141-150.
11. Bradford, M.M. (1976) Anal. Biochem. 72, 248-254.
12. Nagel, S.C.; vom Saal, F.S.; Thayer, C.A.; Dhar, M.G.; Boechler, M.; Welshons, W.V. (1997) Environ. Health Perspect. 105, 70-76.
13. Cox, B.J.; Bunce, N.J. (1999). Anal. Biochem. 267, 357-365.
14. Gaido, K.W.; Leonard, L.S.; Lovel, S.; Gould, J.C.; Babai, D.; Portier, C.J.; McDonnell, D.P. (1997) Toxicol. Appl. Pharmacol. 143, 205-212.
15. Brzozowski, A.M.; Pike, A.C.W.; Dauter, Z.; Hubbard, R.E.; Bonn, T.; Engström, O.; Öhman, L.; Greene, G.L.; Gustafsson, J-Å.; Carlquist, M. (1997) Nature 389, 753-758.

16. Ekena, K.; Katzenellenbogen, J.A; Katzenellenbogen, B.S. (1998) J. Biol. Chem. 273, 693-699.

17. Schneider, U.A.; Brown, M.M; Logan, R.A.; Millar, L.C.; Bunce, N.J. (1995) Environ. Sci. Technol. 29, 2595-2602.

18. Coldham, N.G.; Sivapathasundaram, S.; McDonnell, D.P.; Connor, C.; Sauer, M.J. (1997) Environ. Health Perspect. 105, 734-742.

19. Fielden, M.R.; Chen, I.; Brock, C.; Safe S.H.; Zacharewski T.R. (1997) Environ. Health Perspect. 105, 1238–1248.

20. White, R.; Lees, J.A.; Needham, M.; Ham, J.; Parker, M. (1987) Mol. Endocrinol. 1, 735-744.

21. Shelby, M.D.; Newbold, R.R.; Tully, D.B.; Chae, K.; Davis, V.L. (1996) Environ. Health. Perspect. 104, 1296-1300.

Chapter 3

Analytical Challenges of Environmental Endocrine Disruptor Monitoring

J. M. Van Emon[1], C. L. Gerlach[2], and K.-L. Bowman[1]

[1]**U.S. Environmental Protection Agency, National Exposure Research Laboratory, Human Exposure Research Branch, P.O. Box 93478, Las Vegas, NV 89193–3478**
[2]**Lockheed Martin Services Group, 980 Kelly Johnson Drive, Las Vegas, NV 89119**

Reported increases in the incidence of endocrine-related conditions have led to speculation about environmental causes. Environmental scientists are focusing increased research effort into understanding the mechanisms by which endocrine disruptors affect human and ecological health. Fundamental to this understanding is the ability to characterize and quantify these types of chemical compounds. Immunoassays and tandem analytical techniques, that team chromatography with sensitive immunochemical methods, enable bioanalysis of compounds in complex matrices. Examples of immunochemical methods with applications to environmental monitoring and human exposure assessment are given.

Environmental monitoring and corresponding analytical procedures are evolving to meet a growing number of applications in human and ecological exposure assessments. Recent interest in compounds that are known to or suspected of interfering with the endocrine system present a challenge for widespread monitoring programs and for analytical methods that can generate reliable data for minimum cost. As we describe here, many of these methods already exist but will need to be modified for the demands of environmental endocrine disruptor monitoring studies. Other methods are still in development and may be integrated into monitoring studies as soon as their reliability is established.

Background

In the 1950s, evidence suggested that the pesticide dichlorodiphenyltrichloroethane (DDT), a manmade chemical, was estrogenic. That is, although not a hormone, the compound was able to bind estrogen receptors causing a hormonal effect. Early work in the 1960s further implicated DDT to eggshell thinning, and breeding failures and deformations in birds. The publication of Rachel Carson's Silent Spring in 1962 further raised the issue of environmental effects of synthetic chemicals to the attention of the general public.

In 1970, the U.S. Environmental Protection Agency (EPA) was founded to help safeguard human health and the environment. The use of DDT was banned in the U.S. by the EPA in 1972. However, it is a persistent compound, and more than 25 years later, there are considerable concentrations in some areas. Through the 1980s, more evidence linked some pesticides, plasticizers, and industrial chemicals to endocrine disruption. These chemicals, known as endocrine disrupting compounds (EDCs), and their effects are being studied by various agencies and universities in the U.S. and around the world. Research into EDCs has its roots in several notable achievements in chemistry and biology.

Endocrine System

The endocrine system is actually many glands (i.e., ovaries, testicles, pancreas, adrenal, thyroid, and parathyroid) that have complex interactions and interdependencies. These glands release chemical messengers called hormones into the bloodstream that direct several body functions. The effects of these compounds are spread over time and frequently result in a series of reactions with long sustained effects. For example, ovaries release estrogen, which is necessary for growth, development, and finally reproduction. The thyroid influences metabolism and brain development, and the pituitary controls other glands in the endocrine system. Hormones are directed chemical messengers that exert their effect at very low concentrations. Thus, the detection of these compounds or hormone-like compounds at biologically relevant concentrations require very sensitive analytical methods. An interesting phenomena of the endocrine system is that in many cases it does not appear to be species specific. For example, there are only slight differences among insulin (a product of the pancreas) in humans, horses and cows. Thus, a compound that effects one endocrine system in one species may exert an effect in other species. Frequently in discussions of EDCs only a generic reference is made to the endocrine system.

Characteristics of Endocrine Disrupting Compounds. Given the functional operation of the endocrine system, several compounds can be called EDCs dependent upon dose, frequency and time of exposure. Thus, natural estrogens found in animals (estradiol) and plant phytoestrogens (coumestrol), as well as synthetic estrogens (ethinylestradiol), and some other synthetic chemicals such as diethylstilbestrol (DES), DDT, and polychlorinated biphenyls (PCBs), can all adversely impact the endocrine system dependent upon the time and length of exposure. Individually or in concert, these compounds may present potential threats to human health and ecological well-being. They also present an interesting analytical challenge since there are more than 80,000 candidate compounds for potential testing, including, pesticides, ingredients found in cosmetics, food additives and chemicals listed in the Toxic Substances Control Act (TSCA) inventory (*1*). Clearly, there is a need to establish screening protocols for human and ecological systems.

Analytical Methods

The analytical challenge is to identify, confirm the effects of, and monitor the compounds suspected of impacting the endocrine system. However, there are no common structural characteristics among EDCs, making the task of identifying potential EDCs expensive and labor-intensive. Rapid, reliable and sensitive analytical methods are needed to detect suspected EDCs in a wide variety of matrices. Methods ranging from bioassays to sophisticated analytical procedures are all needed.

A triage approach integrating biological activity and analytical methodologies is needed to unravel the complex interactions of EDCs and their effects. The first tier of analysis would be to determine if a compound actually exerts an endocrine effect. Assays to detect estrogenicity by cell proliferation may be an initial screen to determine if a compound can mimic estrogen action. Other tests would be needed to determine other endocrine effects as no one test can screen for the gamut of endocrine disrupting possibilities. Unfortunately, many of these assays still need to be validated and their reliability determined - particularly for mixtures of compounds. Those compounds giving a positive response in this first battery of assays would then be further evaluated in *in vivo* and environmental studies. This second tier would be indepth studies to determine the manifestation of these effects on animals and ecosystems.

Analytical measurements are the foundation for determining the effect of EDCs in the environment and to ultimately formulate appropriate risk management strategies. The third tier would provide analytical data to determine the routes of exposure. The scope of this undertaking is large given the projected number of suspected EDCs. However, the environmental load of these compounds must be determined in addition to determining their ability to adversely impact the endocrine system.

The analysis of EDCs in multiple matrices will require the seemingly conflicting attributes of simplicity and sensitivity. Simplicity is needed because of the thousands of samples that will need to be analyzed in laboratories and in field locations throughout the country. Sensitivity is critical because of the expected low concentrations of most analytes and the possible subtle interactions of compound mixtures. Immunoassay methods and immunoaffinity chromatography, two analytical techniques with high sensitivity, may play important roles in the study of EDCs.

Immunoassays. Screening methods should be able to detect a broad range of EDCs rapidly and reliably in a high throughput mode. Immunochemical methods, particularly immunoassays, have been used successfully for over a decade in the environmental analysis of pesticides (*2*, *3*). The general characteristics of immunassays make them suitable for supporting EDC studies. They are inexpensive, quantitative procedures that can accommodate large sample loads. Large complex molecules, and small water soluble compounds can be easily detected in both biological and environmental matrices. Many formats are field portable, others can be coupled with instrumental laboratory methods. Immunoassays have been developed for dozens of compounds of environmental concern, including PCBs (*4*), drugs (*5*), and even metals (*6*). Immunoassays have been successfully used to analyze environmental matrices (air, water, soil, food) and biological matrices (human body fluids). Table 1 gives a listing of compounds suspected of being EDCs for which immunoassays have been reported. These existing methods may need to be adapted for particular applications depending upon detection level requirements, matrix and intended use of the data. The partial listing illustrates the wide range of compounds applicable to antibody-based methods. Although antibodies are biologically-derived reagents, immunoassays are physical assays based on the law of mass action. The development steps of an imunoassay can be correlated to those for a gas chromatography (GC) procedure (Table 2). As with all analytical procedures, to obtain reliable data a random sampling of positive and negative responses should be confirmed.

Polychlorinated Biphenyl Analysis. The commercial production and use of PCBs was restricted by the EPA in1977, but due to their persistence in the environment, they still pose a problem in water, sediment, and soil. Thus, exposures to these compounds still occur throughout the environment. PCBs are also suspected of causing endocrine disruption. Air, subsurface soil, and superficial dust samples containing PCBs and other contaminants were collected from an EPA National Priorities List landfill for toxicity testing (*9*). These environmental samples showed endocrine disrupting effects in several bioassays. The quantitative analysis of PCBs in environmental samples is complicated by several factors. There are 209 possible PCB congeners (*10*), only a few of which (notably congeners 77, 126, and 169) are likely to exhibit endocrine disrupting effects. These three congeners are similar in structure to 2,3,7,8-tetrachlorodibenzo-*p*-dioxin (TCDD) and they have relatively high estrogen receptor binding affinities *in vitro* (*9*). They have also been demonstrated to cause uterine weight increase *in vivo* (*9*). The low levels of these congeners in commercially manufactured PCB (Aroclor) range from trace levels to 1.69% (*11*). The low molar percentages make analysis difficult and highlight the need for effective sample cleanup. Even with careful sample preparation, coelution of the minor component of toxic congeners with a major congener may present a problem in analysis.

A quantitative enzyme-linked immunosorbent assay (ELISA) for the determination of PCBs in environmental soil and sediment samples was developed with a linear range of 50-1333 ng/g soil or sediment. Environmental soil samples from an EPA-listed Superfund site were analyzed by GC and by a quantitative ELISA (*12*). For GC analysis the samples underwent a Soxhlet extraction using SW-846 Method 3540A (*13*). For the ELISA, all samples were extracted by shaking with methanol, an abbreviated procedure recommended for commercially-available field test kits. Subsets of the methanolic extracts were further processed using a Soxhlet extraction modified after SW-846 Method 3540A, and a supercritical fluid extraction (SFE) with CO_2. The ELISA results of the methanolic shake extracts were frequently low in comparison to the GC method. However, when the ELISA was the determinative step following extraction by Soxhlet or SFE, the results were essentially identical to GC. Additional soil samples were collected from another contaminated site for ELISA and GC

Table 1. Immunoassay Methods Have Been Reported for These Known and Suspected Endocrine Disruptors (*7*).

2,4-D	Carbaryl	Methomyl
Alachlor	Chlordane	Metribuzin
Aldicarb	DDT	PCBs
Atrazine	Dieldrin	Parathion
Benomyl	Endosulfan	Pentachlorophenol
Cadmium	Lindane	Synthetic Pyrethroids

Table 2. Parallel Processes Between Gas Chromatography and Immunochemical Methods[a].

Gas Chromatography	Immunochemical Methods
Column coating	Antibody selectivity
Column length	Assay format (test tube, microtiter plate)
Number of theoretical plates	Antibody dilution
Column temperature	Incubation temperature
Gases	Buffers
Detector	Detector (Primary: specific antibody) (Secondary: enzymes, radioactivity)
Samples processed serially	Samples processed in parallel
Quantitate from standard curve	Quantitate from standard curve

[a]Reprinted and adapted with permission from reference *8*.

determination. For these samples, the ELISA results for the methanolic shake extracts were biased high when compared to GC. The high-biased data suggested that an interferent was present in the crude, darkly-colored sample extracts. A subset of samples was extracted with an SFE procedure optimized for PCBs which produced cleaner extracts. The SFE-ELISA approach yielded data equivalent to the GC (*12*). The abbreviated sample procedures frequently encountered with field methods are not always reliable and may give rise to false negative or false positive results. Clearly the evaluation of methods must consider all steps in the analytical protocol, especially when introducing new methods for complex problems.

The above ELISA was successfully transferred into another laboratory for additional PCB soil analysis (*14*). The test samples were obtained from the Allied Paper/Portage Kalamazoo River Superfund site. Samples were analyzed by ELISA and GC following a Soxhlet or sonic extraction with methanol. The Soxhlet extraction was found to be more efficient than the sonic extraction for highly contaminated soil samples. A correlation of 0.91 was obtained for the soil samples (n=41). This comparative study also indicates effective sample preparation is a major factor in the analysis of PCBs. The ELISA is being adapted to analyze house dust samples for the determination of indoor exposures.

Pesticide Analysis. The tandem technique of SFE-ELISA has also been used to detect pesticides in food matrices (*15*). Recent studies have utilized spiked baby foods and samples from the U.S. Food and Drug Administration Total Diet Study (TDS). Samples were extracted using supercritical CO_2 and analyzed by various ELISAs. Nine pesticides (alachlor, aldicarb, atrazine, carbaryl, carbendazim, carbofuran, cyanazine, 2,4-D, and metolachlor) were spiked into TDS composite samples. Average recoveries for a 2 ng/g spike ranged from 72% to 108% with the exception of carbendazim (10%), cyanazine (30%), and 2,4-D (ND). This study suggests the feasibility of developing a suite of SFE procedures for extracting groups of pesticides. The off-line coupling with ELISA further diminishes the use of organic solvents, many of which are EDC candidates.

The breadth of immunoassay technology is seen in its applications to various analytes and matrices. The presence and bioavailability of suspected endocrine disrupting pesticides on vegetation may be of concern, particularly for topical residues that are easily dislodged. Exposures from inhalation, dermal absorption or even ingestion can occur due to such foliar dislodgeable residues (FDRs). An ELISA for chlorpyrifos was developed for the analysis of leaf washings from apple, tomato, nectarine, and cucumber foliage (*16*). Leaf punch disks were removed, washed with a dilute surfactant (sodium dioctylsulfosuccinate) and analyzed by ELISA. Control and treated test groups were confirmed by HPLC. The two methods were found to be in close agreement. The ELISA method of determining dislodgeable residues is being extended to determine indoor exposures such as from carpeting.

Immunoaffinity Chromatography. Immunoaffinity chromatography (IAC) has been shown to be an effective cleanup procedure for immunoassays and instrumental analysis. Both immunoassay and immunoaffinity chromatography have been reported for the direct detection of hormonal changes in serum (*17*). A selective antibody coupled to a solid support enables the extraction of a specific target analyte from a sample matrix. Extraneous material elutes through the column while the analyte remains bound to the immobilized antibody. A change in the mobile phase enables the elution of the analyte from the column. The immunopurified extract is usually clean enough for detection and quantitation. High resolution and selectivity have been achieved by coupling IAC to high performance liquid chromatography (HPLC) and/or mass spectrometry (MS). A tandem analytical technique of IAC-HPLC-MS for carbendazim with a detection range of 0.025 to100 ng/mL has been reported (*18*). Variations of this method have been used for detecting anabolic steroids in bile and urine (*19*).

Capillary electrophoresis (CE) is an extremely sensitive separation technology capable of separating enantiomers. CE can be coupled to immunochemical methods for selective sample preconcentration of up to 1000-fold (*20*). This capability may be needed for analyzing EDCs in biological systems, as many hormone-like compounds may exert their effects at or below detection levels of many current analytical techniques.

Human Exposure Assessment

Humans are exposed to relatively high levels of natural EDCs compared to xenobiotic EDCs. To determine if a compound is indeed an EDC, more information is needed on dietary intakes, target organ exposures, mechanism of action, and the interactive effect of mixtures. These and other factors will be crucial to the successful protection of public health and the environment. Compounds that can act on the endocrine system may do so at extremely low levels. It is difficult to extrapolate from a high dose scenario to a low dose model as is frequently done in predicting cancer rates or acute effects. Thus, highly sensitive methods are needed to address the issue of EDCs.

Successful risk management is based on a correct risk conclusion - is a contaminant hazardous and at what levels. Risk conclusions are frequently made on the basis of exposure assessments which in turn, consider the dose response, i.e., the relationship between increased exposure and greater risk. Dose response curves for many suspected EDCs are interestingly shaped, with lower concentrations yielding greater response. The reason for this is not yet understood.

The foundation of the risk characterization process is analytical methodology. All other assumptions and conclusions rely on our ability to separate and detect analytes correctly and consistently. Immunochemical approaches are suitable to the analysis of several EDCs based on target analytes, matrices of interest, and detection level requirements (*21, 22*).

Multidisciplinary and Multiagency Approach. It will be necessary to analyze EDCs in biological fluids (e.g., serum, urine), soil and sediment, groundwater and drinking water, and food and plant material. To this end, several organizations in the U.S. are cooperating by specializing in their own areas of expertise (Table 3). The organizations

Table 3. U.S. Entities Interested in Endocrine Disruptor Screening.

Title	Area of Interest
Environmental Protection Agency	Human and Ecological Health
Department of the Interior	Wildlife Management
Food and Drug Administration	Phytoestrogens
Centers for Disease Control	Epidemiological Studies
Agency for Toxic substances and Disease Registry	Community Impact
National Cancer Institute	Human Health Issues
National Oceanic and Atmospheric Administration	Fisheries and Coastal Ecosystems
National Science Foundation	Support EDC Research
Department of Agriculture	Residues in Meat and Poultry
Department of Defense	Military Operations
National Institute for Occupational Safety and Health	Worker Exposure Issues

are supported by regulations such as the Safe Drinking Water Act (SDWA), the Federal Insecticide, Fungicide, and Rodenticide Act (FIFRA), the Toxic Substances Control Act (TSCA), and the Food Quality Protection Act (FQPA).

In the U.S., a major planning strategy is being developed by the EPA's Endocrine Disruptor Screening and Testing Advisory Committee (EDSTAC). This task force was formed in 1996 as a result of new and amended environmental legislation which in part mandated an increase surveillance of EDCs in natural settings, the effect of EDCs on fish and wildlife, and a concomitant risk assessment of their impact on ecological and human health. The regulatory mandates are under the Federal Food, Drug and Cosmetics Act (FFDCA) as amended by the FQPA. EDSTAC provides advice and council to EPA on screening strategies.

EDSTAC's initial plan is limited to evaluating effects on estrogen, androgen, and thyroid hormonal systems. Further work may include other effects as well as the interaction of mixtures. Thus, the challenge to EDSTAC is enormous and complex: to establish a program for evaluating the toxicity of thousands of chemical compounds and to implement this program into a feasible national process. In August 1998, EDSTAC met its first deadline by publishing and circulating a final draft strategy for the screening and testing of suspect EDCs. The implementation of the plan is Congressionally mandated to be accomplished by August 1999. A report to Congress in August 2000 will complete the EDSTAC mandate.

From the more than 80,000 candidate compounds, an approximate initial number of 15,000 will be evaluated. It is anticipated that 1,000 of these compounds will be further studied. Candidate compounds, such as pesticides and detergent additives will be evaluated first due to their exposure potential. Evaluation of polymers and other compounds with low bioavailability will be initially delayed. It is possible that relatively few (30-50) compounds will be the principal EDCs and that addressing the problem will be simpler than designing the study.

Notice

The U.S. Environmental Protection Agency (EPA), through its Office of Research and Development (ORD), funded and managed the research described here under Contract 68-C5-0091 to Lockheed Martin Services Group. It has been subjected to the Agency's peer review and has been approved for publication. Mention of trade names or commercial products does not constitute endorsement or recommendation by the EPA for use.

Literature Cited

1. *Endocrine Disruptor Screening and Testing Advisory Committee (EDSTAC)*; final report to the U.S. Environmental Protection Agency, Office of Pollution, Prevention, and Toxic Substances; U.S. Government Printing Office: Washington, DC, 1998.

2. Van Emon, J. M.; Seiber, J. N.; Hammock, B. In *Analytical Methods for Pesticides and Plant Growth Regulators: Advanced Analytical Techniques*; Sherma, J., Ed; Academic Press: New York, NY, 1989, Vol. XVII; Chapter 7, pp 217-263.

3. Van Emon, J. M.; Lopez-Avila, V. *Anal. Chem.* **1992**, *64*, 78A-88A.

4. Van Emon, J. M.; Gerlach, C. L.; Johnson, J. C. In *Current Issues in Regulatory Chemistry*; MacNeil, J. D.; Kay, J. F.; O'Rangers, J. J., Eds; AOAC International: Gaithersburg, MD, in press.

5. Haasnoot, W.; Stouten, P.; Lommen, A.; Cazemier, G.; Hooijerink, H.; Schilt, R. *Analyst* **1994**, *119*, 2675-2680.

6. Blake, D. A.; Blake, R. C., II; Khosraviani, M.; Pavlov, A. R. *Anal. Chim. Acta* **1998**, *367*, 13-19.

7. Colborn, T. *Arctic Res. U S* **1994**, *8*, 220-231.

8. Van Emon, J. M.; Gerlach, C. L. *Inside Lab. Manag.* **1997**, *March*, 31-33.

9. Li, M.-H.; Hansen, L. G. *Environ. Health. Persp.* **1996**, *104(7)*, 712-722.

10. Albro, P. W.; Parker, C. E. *J. Chromatogr.* **1979**, *169*, 161-166.

11. Albro. P. W.; Corbet, J. T.; Schoeder, J. L. *J. Chromatogr.* **1981**, *205*, 103-111.

12. Johnson, J. C.; Van Emon, J. M. *Anal. Chem.* **1996**, *68(1)*, 162-169.

13. *Method 8080-Organochlorine Pesticides and PCBs*; Manual SW-846: 1986; U.S. Environmental Protection Agency. Office of Solid Waste and Emergency Response. U.S. Government Printing Office: Washington, DC, 1986.

14. Chuang, J. C.; Miller, L. S.; Davis, D. B.; Peven, C. S.; Johnson, J. C.; Van Emon, J. M. *Anal. Chim. Acta* **1998**, *376*, 67-75.

15. King, J. W.; Nam, K.-S. In *Immunoassays for Residue Analysis Food Safety*; Beier, R. C.; Stanker, L. H., Eds; ACS Symposium Series 621; American Chemical Society: Washington, DC, 1996; Chapter 34, pp 422-438.

16. Van Emon, J. M.; Gerlach, C. L.; Reed, A. W.; Hardwick, B. C. *Food Technol. Biotechnol.* **1998**, *36(2)*, 119-124.

17. Palmer, D. A.; Evans, M.; Miller, J. N.; French, M. T. *Analyst* **1994**, *119*, 943-947.

18. Thomas, D. H.; Lopez-Avila, V.; Betowski, L. D.; Van Emon, J. M. *J. Chromatogr. A* **1996**, *724*, 207-217.

19. Fodey, T. L.; Elliott, C. T.; Crooks, S. R. H.; McCaughey, W. J. *Food Agric. Immunol.* **1996**, *8*, 157-167.

20. Cole, L. J.; Kennedy, R. T. *Electrophoresis* **1995**, *16*, 549-556.

21. Van Emon, J. M.; Gerlach, C. L. *J. Microbiol. Meth.* **1998**, *32*, 121-131.

22. Van Emon, J. M.; Gerlach, C. L.; Bowman, K. *J. Chromatogr. B* **1998**, *715*, 211-228.

Chapter 4

Determination of Suspected Endocrine Disruptors in Foods and Food Packaging

Timothy P. McNeal[1], John E. Biles[1], Timothy H. Begley[1], John C. Craun[2], Marvin L. Hopper[2], and Chris A. Sack[2]

[1]**Food and Drug Administration, 200 C Street, SW, Washington, DC 20204**
[2]**Food and Drug Administration, 11630 West 80th Street, Lenexa, KS 66214**

Bisphenol-A (BPA), nonyl phenol (NP), and certain phthalate esters are common industrial chemicals employed in a variety of ways to manufacture, stabilize or modify the characteristics and performance of polymers, including those that are approved for use in food packaging. BPA is used as the starting material for the synthesis of polycarbonate (PC) plastics used to make baby bottles and reusable water carboys. BPA is also used to make epoxy adhesives and can coatings. The major source of NP residues in food packaging is the oxidation of tris-nonylphenyl phosphite (TNPP), an antioxidant/antiozonant, added to polymeric materials such as polyvinylchloride (PVC), polyolefins, and acrylics. *Ortho-* phthalate esters such as dibutyl, butylbenzyl, and di-2-ethylhexyl phthalate have been widely used as plasticizers in PVC and are also used in adhesives and printing inks. Phthalate esters are commonly found in the environment and are generally considered ubiquitous.

Recently, preliminary data reported in the literature, primarily from *in vitro* studies, suggest that some of these chemicals may have weak estrogenic activity, although their effects in animals and humans are far from clear. We will refer to BPA, NP and phthalates as suspected endocrine disruptors (EDs). Because these suspected ED's are present as additives or residues in food-contact materials, they can be expected to migrate to some foods in very low amounts. Larger amounts of migration can be expected from polymers exposed to food at elevated temperatures, i.e., heat-processed foods. The levels of these chemicals that migrate into foods are of interest to the Food and Drug Administration (FDA) because estimates of dietary exposure are needed in order to assess any potential health significance.

FDA's indirect additives laboratory conducted a limited survey of food packaging for suspected EDs. A portion of each package was extracted with hexane or methanol, and suspected EDs determined by HPLC with fluorescence detection or gas chromatography with mass selective detection. The methodology was straightforward and yielded many sources of suspected EDs. BPA residues were detected in all PC items tested, and trace amounts in many epoxy-based enamels, including infant formula cans. NP residues were found in many food contact plastics from rigid polystyrene cold drink cups to laminated films. Traces of the phthalate esters were predominantly found in colored laminated films and are thought to be associated with color or ink formulations.

Analytical techniques for the determination of suspected ED residues in foods are matrix- and analyte- dependant. BPA residues migrating from PC food contact plastics to food simulating solvents and infant formulas were determined after dilution with mobile phase and HPLC analysis with fluorescence detection. NP and BPA residues from can enamels and jar lid gaskets were determined in many aqueous foods by using solid phase extraction with a porous polymer and then analysis with either capillary GC with mass selective detection, or HPLC with fluorescence detection. Another study, an FDA field assignment on phthalates in infant formulas, was conducted by FDA's Total Diet Laboratory in Lenexa, KS. Composites representing 81 different powdered and liquid infant formula samples were analyzed for dibutyl, butylbenzyl, and di-2-ethylhexyl phthalates. The samples were subjected to multi-residue *FDA Pesticide Analytical Manual* extraction protocols and the phthalates determined by capillary GC-MS. In general, either non-detectable or low μg/kg amounts of the suspected EDs were found in the foods analyzed.

On the basis of *in vitro* testing that demonstrated weak estrogenic activity, many industrial chemicals are suspected to adversely affect the endocrine systems of animals. Some of the suspected endocrine disruptor (ED) chemicals are bisphenol-A (BPA), nonyl phenol (NP), and the di-n-butyl (DBP), butyl benzyl (BBP), and di-2-ethylhexyl (DEHP) *ortho*-phthalate esters (*1-12*). All of these chemicals are used to make or improve the performance of plastics and coatings. Also, these chemicals are permitted for use in food packaging by the Food and Drug Administration (FDA) (*13*). Although no direct link to endocrine abnormalities in humans has been established, prudence would dictate that reasonable steps be taken to evaluate man's dietary exposure to these chemicals.

Bisphenol-A is combined with epichlorohydrin to form the diglycidyl ether of bisphenol-A (DGEBP), the major building block of most epoxy enamels, used to make can liners and components of non-stick surfaces of metallic cookware. Bisphenol-A is used to make adhesives, phenolic resins used as can liners, and as a co-monomer for high impact polycarbonate (PC) resins used to make repeat-use food contact items such as water carboys, food processors, and baby bottles. Low levels of residual BPA are trapped in the enamels and PC resins due to incomplete polymerization, and these residues may migrate into foods.

Nonyl phenol derivatives are approved for use in adhesives, paper, and resins and coatings for food packaging. Nonyl phenol is also used for the preparation of ionic and non-ionic surfactants, the majority of which are non-ionic ethoxylated nonyl phenols that are approved for use in food contact materials. Also, NP is used to make the antiozonant, tris-nonyl phenyl phosphite (TNPP) which may be used in polyvinyl-chloride (PVC), polyolefins and acrylic polymers. Ethoxylated nonyl phenols are relatively stable in food packaging under approved conditions of use, but TNPP is not. It readily oxidizes, as it is intended to do, forming NP, which readily migrates into foods.

Ortho-phthalate esters had been used for many years as primary plasticizers in PVC film and tubing used for food contact applications. Also, the *ortho*-phthalate esters are approved for use in adhesives. The US industry has stated that DBP, BBP and DEHP are no longer used in PVC for food contact use, and our surveillance studies confirm this.

Chemical Residues in Food Packaging

Our laboratory routinely performs screenings to determine chemicals contained in food packaging as well as the presence of suspected ED's and chemical contaminants. We analyze food packages purchased in the Washington, DC area. Also, every 3 months as part of FDA's total diet study (TDS), the Kansas City laboratory sends to us selected packages such as cheese and cookie packages for screening. With the analysis of these packages, estimates of the concentrations of various chemicals in the packages are obtained with either external or internal standardization. Also, analysis of these packages gives us data on chemical additive use in food packaging across the country, and helps us keep up with food packaging trends.

Typically, the analysis for residual chemicals in food packaging involves solvent extraction or dissolution of the package. For example, residual BPA in PC containers is determined by dissolving the polymer container in methylene chloride, the polymer reprecipitated with excess methanol, and the combined solvent extract is filtered, diluted with mobile phase, and residual BPA determined by HPLC with fluorescence detection (*14*). The concentration of BPA in can enamels is difficult to determine because the mass of the enamels is unknown. At best, BPA can be extracted from a can enamel with chloroform, the extract concentrated, and the presence of BPA confirmed by GC-MS. Once the presence of BPA is confirmed in the package, then further analysis of the food may be warranted.

Nonyl phenol derivatives and the *ortho*-phthalates are approved for use in many different types of food packages. Therefore, the potential exists for these chemicals to be in many different foods. However, it can not be assumed that residues of these chemicals will be found in all packages for which they are approved for use.

Hence, many food packages have to be screened for the presence of these chemicals. The procedure is straightforward. The chemicals are soluble in hexane and easily extracted from the different packages. Also, the analytes are amenable to multi-residue analysis with capillary GC. This allows for the determination of multiple analytes in one analysis.

Recently, low levels of DEHP were reported migrating to cheddar cheese from plastic food packaging (*16*). Limited data from our laboratory suggest their source to be laminating adhesives, print inks used to decorate packages or describe the contents, and pressure sensitive adhesives used with package labels. However, even with these limited uses in food packaging, years of phthalate use in consumer items such as vinyl upholstery and floor tiles make them ubiquitous in the environment.

Chemical Residues in Foods

Krishnan demonstrated that BPA migrated from flasks made with PC into biological culture media during autoclaving (*4*). In the case of food packaging, BPA migration into water bottled in PC carboys and infant formula prepared in PC baby bottles were investigated (*14*). Brotons found BPA residues in vegetables processed in cans lined with epoxy or modified epoxy food contact surfaces (*9*). If BPA migrates from can liners into vegetables, how much migrates into canned infant formula? Our initial focus was to answer this question. Then we investigated BPA migration into other foods.

Nonyl phenol migrates into foods, whether its presence is the result of the oxidation of TNPP or as a contaminant in TNPP or ethoxylated nonyl phenols. Of primary interest, is NP migration from vinyl jar lid gaskets (PVC base) into any type of bottled baby food. Also, NP migration from PVC film is of high interest because these films have been shown to contain the highest levels of NP. Nonyl phenol residues extracted from laminated olefin films, rigid polystyrene, and some acrylics are much lower than the levels observed in jar lid gaskets and PVC film. Therefore, the corresponding migration levels from these other plastics would be much lower, and more difficult to determine.

In 1996, the United Kingdom's (UK) Ministry of Agriculture, Fisheries and Food (MAFF) released a food surveillance report on phthalates in infant formula in the UK (*15*). They reported up to 1 µg/g DEHP and from 0.08 to 0.40 µg/g DBP in all of 12 powdered infant formula samples tested. In response to this report, FDA's Center for Food Safety and Applied Nutrition (CFSAN) issued a phthalates in infant formula field assignment. The task was to determine the levels of phthalates in infant formula sold in the United States, and the analyses were performed by FDA's Kansas City District Laboratory.

Determinations for these suspected ED's in a variety of different foods and food simulating solvents require different analytical procedures because of the varied matrices and polarities of the analytes. Also, different means of detection will have to be used to gain maximum selectivity and sensitivity for the analytes. However, because of the variety of procedures employed and the space restrictions of this publication, experimental procedures will be generalized and this paper will place emphasis on the results rather than the intricacies of the varied methods use to obtain the data.

Experimental

Analytes and Standard Solutions. Nonyl phenol, technical (84852-15-3), 1,2-di-n-butyl phthalate, 99+% (84-74-2), bisphenol-A, 99+% (80-05-7), 1,2-benzyl butyl phthalate, 98 %, (85-68-7), 1,2-dicyclohexyl phthalate (DCHP), 99%, (84-61-7), and 1,2-di-2-ethylhexyl phthalate, 99%, (117-81-7). All chemicals were purchased from Aldrich Chemicals, Milwaukee, WI.

Standards for GC-MSD Analysis of Suspected ED's in Food Packaging. Prepare DCHP internal standard (for phthalates) stock solution (ca. 2000 μg/mL) in hexane. Except for BPA, prepare mixed stock standard (ca. 2000 μg/mL) in hexane. Prepare BPA standard in chloroform. Prepare separate internal and mixed working standards (ca. 40 μg/mL) in hexane.

Standards for GC-MSD Analysis of Suspected ED's in Foods and Food Simulants. Prepare ca. 500-1000 μg/mL stock solutions. Then prepare working standards in chloroform in the 1–10 μg/mL range.

Standards for Phthalates in Infant Formula. Prepare mixed phthalate stock standard (ca. 1000 μg/mL) in iso-octane. Dilute stock standard with 10% acetone in iso-octane to obtain 4 working standards ranging from 0.1 to 2.0 μg/mL.

Instrumentation. GC-MS-DS for All Analyses Other Than Phthalates in Infant Formula. Hewlett-Packard (HP) 5890A gas chromatograph with HP 7673 automated liquid sampler (ALS), spilt-splitless injector, and capillary direct interface to an HP 5970B mass selective detector (MSD); column, 30 m x .25 mm Rtx-5MS FSOT capillary (5 % diphenyl – 95 % dimethyl polysiloxane) with 0.25 μm bonded film; helium carrier at 1 mL/min (25 °C); temperatures (°C), injector 270, interface 280, oven program, 2 min at 100, 10 °/min to 280, hold 10 min; total time, 30 min; 2 μL injection, split vent opened after 1.5 min. Retention times of analytes, NP, DBP, BPA, BBP, DCHP and DEHP ca. 10.5 - 12.0, 13.9, 16.0, 17.6, 19.0 and 19.2 min respectively. Instrument control and data acquisition and analysis were accomplished with an HP ChemStation with version 3.2 Pascal operating software. The MSD was optimized by using the Autotune program with perfluorotributylamine as the calibration standard.

MSD Parameters for Confirmation and Quantitation of ED's in Food Packages. The operating parameters were as follows: scan acquisition mode, scan range (m/z) 50 – 450, and the scan rate was ca. 1.1 scans/sec. Confirmation was obtained by matching full mass spectra to spectra of authentic standards with a fit confidence level > 70% and retention times within a window of 20 seconds of those from authentic standards. Quantitation was based on the integrated response of the quantitation ion for each compound, (m/z) 149 for phthalates, 135 for NP, and 213 for BPA. Calibration was accomplished by using internal standardization with DCHP for phthalates, and external standardization for NP and BPA.

MSD Parameters for Confirmation and Quantitation of NP and BPA in Foods and Food-Simulating Solvents. Acquisition was in the selected ion monitoring mode. The dwell time was 50 msec and the acquisition rate was 3.7

cycles/sec. The ions (m/z) monitored were 119, 213 and 228 for BPA, and 107, 135 and 220 for NP. With this configuration, confirmation was based on the normalized integrated response ratios (± 20%) of the peaks for the given ions. With our system, the ratios of 0.18:1.00:0.30 were obtained for BPA. For NP, either of 2 different criteria were used;

1) The normalized integrated response for 1 of 3 major peaks, i.e., with our GC-MS system, the peak at 11.1 min the ions (m/z) 107, 135, 220 had corresponding normalized integrated response ratios of 0.16:1.00:0.03. The peak at 11.2 min had normalized integrated response ratios of 0.55:1.00:0.07, and the peak at 11.7 min had normalized integrated response ratios of 0.30:1.00:0.02.
2) The ratios for 3 of 5 of the normalized integrated responses of the chromatographic peaks for the major NP ion, (m/z) 135. With our GC-MS system, these ratios were 1.00:0.56:0.46:0.39: 0.83 for the 5 major chromatographic peaks.

Quantitation was by external calibration and based on the integrated response of (m/z) 213 for BPA, and (m/z) 135 for any one of the major NP isomers.

HPLC Parameters for the Analysis of PC Containers, Foods and Food Simulants for BPA. Follow the protocol outlined by Biles, et.al. (*14, 17*).

GC-MS-DS for Phthalates in Infant Formula. A Varian Star 3400 CX gas chromatograph equipped with a programmable septum-equipped injector, a Varian 8200 automated liquid sampler, and an ion trap mass spectrometer; column, 30 m x .25 mm DB-5 FSOT capillary with 1.0 µm bonded film; 1 mL/min helium carrier gas flow at 150 °C; temperatures (°C), injector program, ramp from 80 ° at 180 °/min to 260°, hold 16 min; oven program, start at 150 °, program at 10 °/min to 260 °, hold 9 min. Injection technique, slow splitless injection (1 µL/sec) of 1 µL as septum-equipped injection with insert packed with 20SE sorbent. Retention times of analytes, DBP, BBP and DEHP ca. 8.0, 11.7 and 13.7 min respectively.

Sample Preparation. Preparation of Food Packages for Analysis. Rinse packages with distilled water to remove food residues and dry with absorbent toweling. Cut pieces of flexible packages representative of total package. Transfer 0.5 g of pieces to a 20 mL headspace vial. Add 5 mL hexane to vial, and for phthalate esters, add internal standard to give ca. 1 µg/g DCHP in the package specimen. Seal each vial with a PTFE septum and place in rotating mixer for overnight extraction at room temperature. For the extraction of jar lid gaskets, add 5 mL hexane to glass container, reset lid to container, and invert container for overnight extraction. After extraction, transfer 1 mL aliquot to ALS vial, seal the vial, and analyze extract by capillary GC with mass selective detection in the scan mode.

Prepare PC containers for HPLC analysis using protocol described by Biles, et.al. (*14*).

Preparation of Foods and Food Simulating Solvents for BPA Determinations. Prepare infant formula and fruit juices for HPLC analysis, and bottled water for GC-MS analysis by using protocols described by Biles, et.al. (*14, 17*). For analysis of canned vegetables except tomatoes, decant liquid, weigh the liquid, and blend liquid at high speed in a Waring blender for 5 min. Quickly transfer

50 - 200 g of homogenate to tarred 250-ml polyethylene centrifuge bottle, and centrifuge for 10 min at 3000 rpm's. Decant and weigh liquid. Proceed with solid phase extraction (SPE) procedure described by Biles, et.al. (*17*). For tomatoes, weigh contents of can, blend the total contents, and centrifuge a 50 – 200 g aliquot of product. Decant, weigh the liquid, and proceed with the SPE procedure. Bring final extract up to 1 mL with chloroform for GC-MSD analysis. For HPLC analysis bring up to 1 mL with mobile phase.

Preparation of Foods and Food Simulating Solvents for NP Determinations. Dilute infant formula aliquots 1:10 with water. Dilute aliquots of other foods 1:1 with water. Proceed with SPE procedure for BPA analysis, except retain the hexane eluate. Bring the hexane eluate to dryness and bring up to 1 mL with chloroform for GC-MSD analysis.

Preparation of Infant Formula for Phthalate Determinations. Composite infant formula units by brand, type (milk or soy base and low or high iron fortification), and form (powder, ready-to-feed, liquid concentrate and ready-to-feed nursette). Prepare each composite for GC- ion trap mass spectrometric analysis by using the following protocols from Volume 1 of the *Pesticide Analytical Manual* (PAM) (*18*). Extract 50 g of composite with 100 mL of acetonitrile for 2 minutes. Centrifuge at 1200 rpm for 5 minutes. Dehydrate acetonitrile extract as outlined in procedure E2 of section 302 with the final extract brought up in hexane rather than acetone. Pre-rinse an activated florisil column with 20 mL hexane. Pass hexane extract through florisil column as outlined in procedure C5 of section 302 except for the following modifications: Elute the phthalates with 50 mL of 1:1 petroleum ether and ethyl ether, concentrate the mixed ether extract to near dryness, add 25 mL hexane through Snyder column, evaporate to near dryness, and bring up to 1.4 mL with hexane. The extract is now ready for GC analysis with ion trap mass spectrometry.

Results and Discussion

Food Packaging Analyses. Packages analyzed in our surveys included many different brands and types of wraps, i.e., PVC film, bread bags, laminated pouches etc... Also, many different jar lid gaskets, plastic cups, and many other packages were analyzed.

The two major sources of exposure to BPA from food packaging are can enamels and PC containers. For epoxy base can enamels, BPA was extracted into chloroform, and high ng/mL levels were determined by HPLC with fluorescence and confirmed by GC-MS. Bisphenol-A concentrations in the enamels could not be determined because the mass of the enamels was unknown. However, BPA levels in PC baby bottles were determined. In a limited survey of baby bottles from 6 different manufacturers, we found residual BPA levels in all bottles, ranging from 7 to 47 μg/g (*14*). Residual BPA levels were not determined in PC water carboys, but the levels are expected to be within the same range observed in the baby bottles because both types of containers are molded from similar resins.

Nonyl phenol can migrate into food from packaging through a number of routes. The major source of exposure is the result of the oxidation of TNPP. In a limited survey, PVC stretch films from different grocery stores used to package meats and vegetables, and different brands of consumer stretch wraps were analyzed for NP. Nonyl phenol was found at levels up to 0.1% in all grocery store PVC stretch wraps and one brand of consumer PVC stretch wrap. Figure 1 represents the GC-MS analysis of the hexane extract of one of the surveyed packages, a PVC produce wrap. The upper portion of the figure is the total ion chromatogram of the hexane soluble semi-volatiles extracted from the wrap. The major peak in the chromatogram is di-2-ethylhexyl adipate, the primary monomeric plasticizer. Also seen in the chromatogram are the numerous NP isomers in the region from 10.5 to 12 min (NP is prepared by the alkylation of phenol with the trimer of propylene. The propylene trimer consists of many isomers). The GC-MS analysis of alkyl phenol standards indicates that the major NP isomers have highly branched nonyl groups. With the described column, most of the major NP isomers elute prior to *para*-n-octyl phenol. The lower portion of figure 1 is the ion chromatogram for m/z 135, the most intense ion of the major NP isomers. The pattern of this ion chromatogram could be regarded as the NP signature. Nonyl phenol residues measured in the PVC wrap were ca. 250 µg/g and arise from the oxidation of TNPP.

Many vinyl jar lid gaskets thought to contain TNPP had NP residues. These included lid gaskets for jars of infant formula, baby food juices, vegetables and meals. Based on the hexane extraction of NP from pieces of the gasket materials, we estimate those levels to be at low µg/g levels. We also found NP residues in some polystyrene-, polyacrylic-, and polyolefin- based food packages such as rigid cold drink cups, blister packages and aseptically sterilized beverage containers at very low µg/g levels.

Both clear and color-laminated cheese packages, and some with adhesive-bound labels, were analyzed as potential sources of phthalate ester contamination. Data that were typical of these analyses are illustrated in figure 2. The upper portion of figure 2 is the total ion chromatogram from the analysis of the hexane extract of a clear colorless plastic laminated cheddar cheese wrapper with a pressure sensitive adhesive label. Even though the packaging was rinsed and dried to remove cheese residues, the major peaks seen in the chromatogram are hydrocarbons, fatty acids and esters from the cheese. Also in the extract, are 9.5 µg/g DBP and 9.1 µg/g DEHP. The 2 lower scans from the total ion chromatogram are the mass spectra representing DBP at 13.9 min (lower left) and DEHP at 19.2 min (lower right).

In a survey, sixteen different cheese packages were analyzed for phthalate residues. Four of the packages were purchased in the Washington, DC area, and the remaining 12 were from 4 TDS collection sites. In all, the packages represented different cheeses purchased from 13 different US cities. Di-n-butyl phthalate was not detected in any of the packages. Butyl benzyl phthalate was found in 3 of the packages at levels from 0.6 to 8.8 µg/g. Di-2-ethylhexyl phthalate was found in all but 1 of the packages at levels from 1 to 22 µg/g. The detection limit for the phthalates in the packages was 0.1 µg/g.

In the next experiment, different portions of selected cheese packages were extracted and analyzed for phthalates. Portions with or without adhesive labels and with or without colored/printed regions were analyzed. We saw DBP levels in regions with labels and colors or printing in three of the packages at levels from 5.8 to 9.5 µg/g, and DEHP levels were 2 to 4 times higher in regions with labels and colors or printing. This strongly suggests that print inks and adhesives are the probable sources of the phthalates.

Food Analyses. Bisphenol-A Determinations. Experiments were conducted to determine the levels of BPA that migrate from reuseable PC baby bottles into infant formula and juices. Upon completion of a migration experiment under realistic sterilization and use conditions, 30 minutes at 100 °C followed by 72 hours under refrigeration, the formula was diluted with water and analyzed by HPLC (*14*). No BPA was found to migrate into formula or juice at a detection limit of ca. 100 ng/mL. The absence of any cleanup and matrix effects accounted for the high detection limit. However, when similar migration experiments were run using food simulating solvents such as distilled water and 10% ethanol, which eliminate background interference's seen in infant formula, traces of BPA migration, equivalent to ca. 2 ng/mL in infant formula, were determined. Although BPA residues measured in PC baby bottles ranged from 7 to 47 µg/g, only ca. 2 ng/mL were suggested to migrate into the infant formula, and these migration levels represent only a very small fraction of the amount of BPA in the PC containers.

Next, BPA migration from 5-gallon PC carboys into water was determined. Water from a bottled water station at our laboratory facility was periodically sampled. The fill date on the carboy was recorded whenever the water was sampled. For the water analysis, a 1 L aliquot of the bottled water was passed through a C-8 SPE cartridge. The BPA was concentrated on the cartridge, easily eluted with a few mL's of chloroform, and readily analyzed by capillary GC with mass selective detection in the selected ion monitoring mode (*14*). The following table summarizes the bottled water data.

Table I. Bisphenol-A Residues Found in Water Bottled in 5-gal Polycarbonate Carboys

Time[a] (weeks)	BPA found (ng/mL)
39	4.7, 4.6
12	0.5, 0.4
3	0.1, 0.2
Distilled water control	<0.1

[a] Elapsed time that water was in carboy prior to analysis.

As expected, the data show that the concentration of BPA in water increased with contact time. Bisphenol-A migration into water from PC carboys ranged from 0.1 ng/mL after 3 weeks of contact to a maximum of 4.7 ng/mL after 39 weeks of contact. The detection limit for these analyses was 0.05 ng/mL and recoveries at the 1 and 3 ng/mL levels ranged from 98 to 105%. Residual BPA was not determined in the polycarbonate water carboys. The levels are expected to be within the same range observed in the baby bottles for the reasons that were previously mentioned.

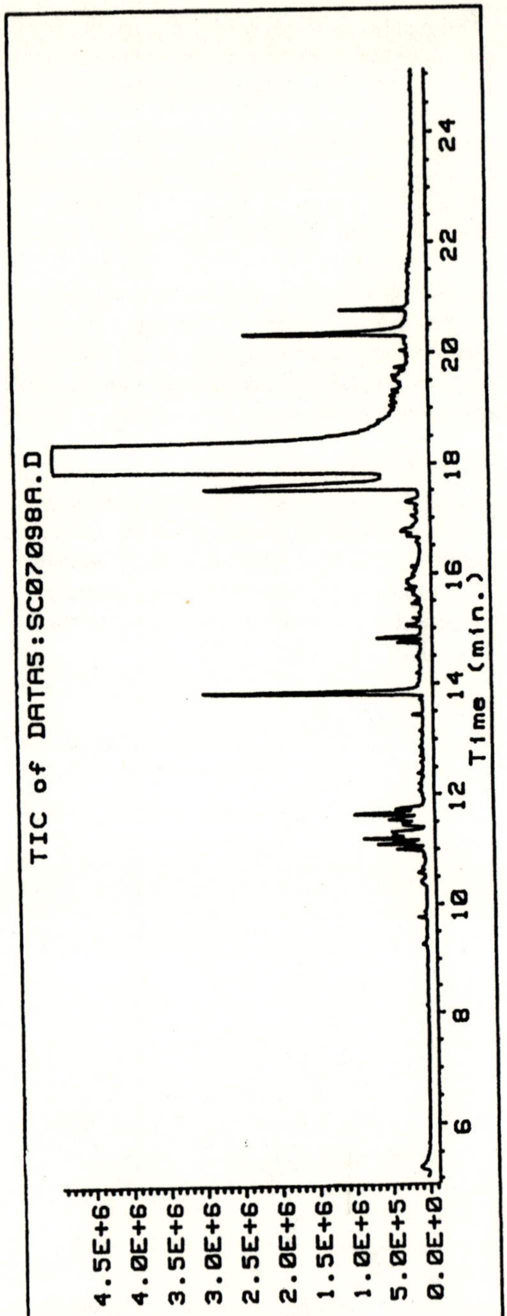

TIC of DATA5:SC07098A.D
Abundance
4.5E+6
4.0E+6
3.5E+6
3.0E+6
2.5E+6
2.0E+6
1.5E+6
1.0E+6
5.0E+5
0.0E+0
6
8
10
12
14
16
18
20
22
24
Time (min.)

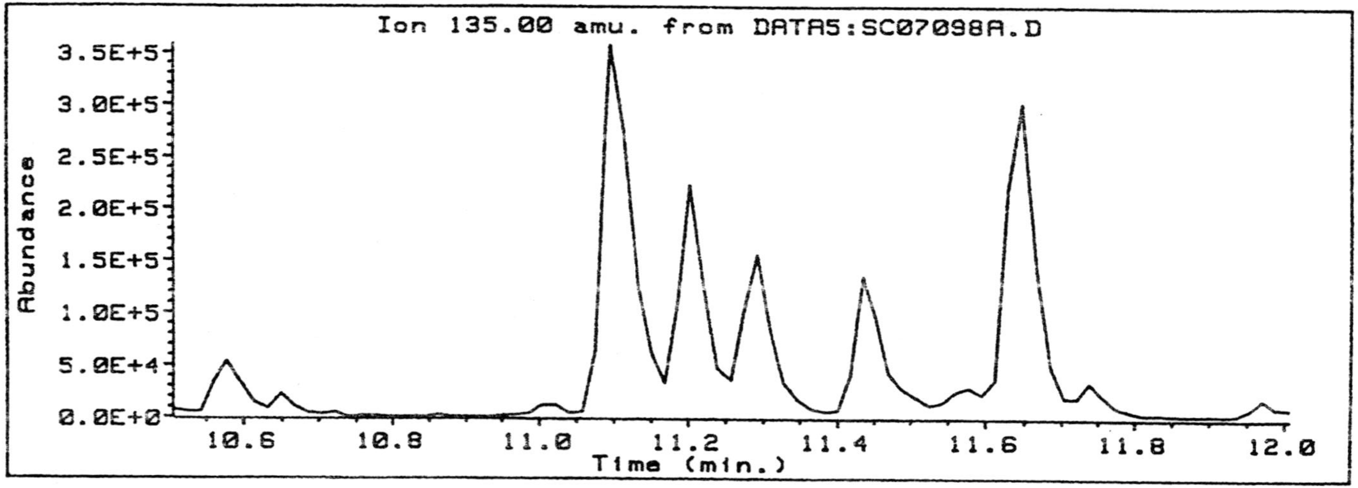

Figure 1. GC-MS analysis of the hexane extract of a PVC produce wrap. The upper figure is the total ion chromatogram of the hexane extractable semi-volatiles. The lower figure is the extracted ion chromatogram for m/z 135, the most abundant ion of the nonyl phenol isomers.

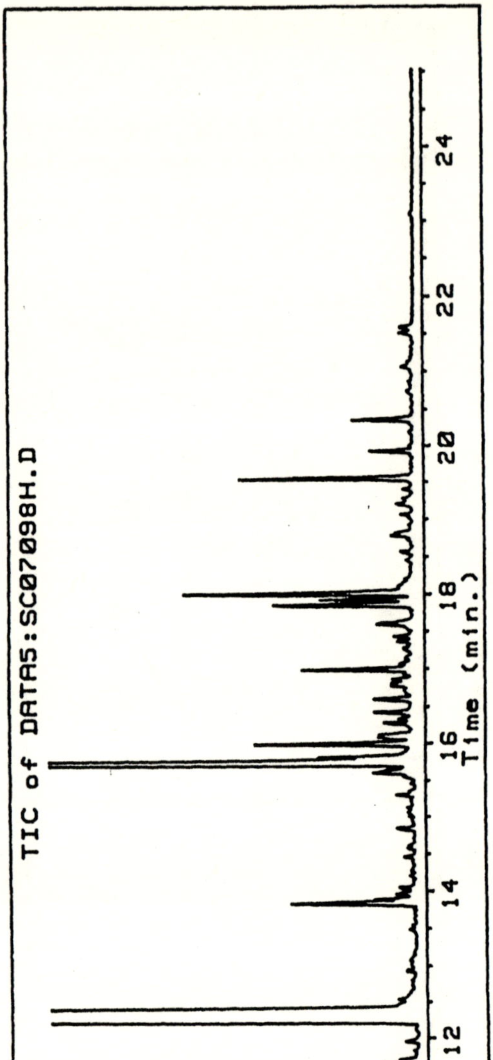
TIC of DATA5:SC07098H.D
Abundance
5.0E+6
4.5E+6
4.0E+6
3.5E+6
3.0E+6
2.5E+6
2.0E+6
1.5E+6
1.0E+6
5.0E+5
0.0E+0
12
14
16
18
20
22
24
Time (min.)

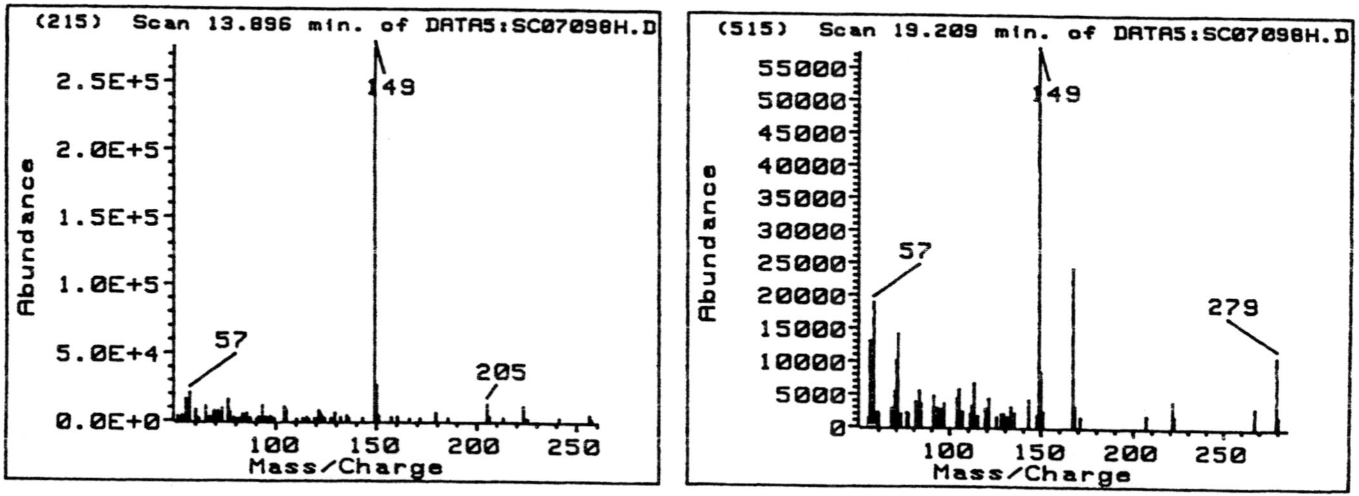

Figure 2. GC-MS analysis of the hexane extract of a cheddar cheese wrapper. The upper figure is the total ion chromatogram of the hexane extractable semi-volatiles. The lower figures are full mass scans from the chromatogram and represent on the left, di-n-butyl phthalate, and on the right di-2-ethylhexyl phthalate.

Therefore, the data show very low parts-per-billion levels of BPA migrating from polycarbonate containers into water, which is also consistent with the migration to infant formula.

The amount of BPA migration from can enamels into infant formula was also determined. Different brands of canned infant formula were purchased in the Washington, DC area and analyzed for BPA. An aliquot of formula was diluted with water, and BPA residues were concentrated on an SPE cartridge loaded with a cross-linked styrene-divinyl benzene solid phase. Bisphenol-A was eluted with a small portion of chloroform, the volume reduced, and brought up in mobile phase. Then the final extract was analyzed for BPA using reversed phase HPLC with fluorescence detection (*17*). Also, selected infant formula extracts were further concentrated, brought up in chloroform, and the presence of BPA in the concentrates confirmed by capillary GC with mass selective detection in the selected ion monitoring mode. Table II summarizes the migration data.

Table II. Bisphenol-A Found in Canned Infant Formula Concentrates

Brand	Can Type[a]	BPA (ng/mL)
A	1	13.2, 12.1
B	1	8.3
C	2	1.5, 1.3
C[b]	2	1.3
E	3	9.5
A	4	5.1
B	4	4.8, 3.6
D	4	0.1, 0.7
D[b]	4	3.9
C[b]	5	4.5

[a] 1 = 2 piece can, epoxy lid, modified epoxy body.
2 = 2 piece can, epoxy lid, vinyl body.
3 = 3 piece can, epoxy ends, modified epoxy sidewall.
4 = 3 piece can, epoxy ends, and sidewall.
5 = 3 piece can, epoxy ends, modified epoxy sidewall, ready-to-feed formula.
[b] Soy based formula.

As with migration from PC containers, the data show that very low parts-per-billion levels of BPA migrate from epoxy base can enamels into infant formula. The data show BPA migration from epoxy can coatings into infant formulas ranged from 0.1 to 13.2 ng/mL. Because all but one of the formulas tested were concentrates, BPA consumed would be no more than 6.6 ng/mL after a 1:1 dilution with water.

Investigations into the migration of BPA to vegetable liquids from epoxy-coated cans were performed. These were canned vegetables purchased in local grocery stores. Except for canned tomatoes, only the aqueous portion of the products was analyzed. Solids were separated by centrifugation, and the liquids subjected to the same SPE procedure used for canned infant formula. Then comparative analysis of the final extracts was performed by using both reversed phase HPLC with fluorescence detection, and capillary GC with mass selective detection in the selected ion monitoring mode. Table III summarizes the amounts of BPA residues found in canned vegetables.

Table III. Bisphenol-A Residues Found in Canned Tomatoes and Vegetable Liquids

Canned Vegetable	Bisphenol-A Found (ng/mL)	
	GC-MSD	HPLC-Fluorescence
Mushrooms, canned	15, 11	12.5, 9.4
Mushrooms, jar	2, 5, 5, 6	not determined
Green beans	15, 16, 12, 9	6.1
Mixed vegetables	2.6, 6.2, 12.2	38.7, 9.1
Artichoke hearts	11, 5, 4, 8	not determined
Tomatoes, can 1	21, 21, 21	not determined
Tomatoes, can 2[a]	21, 18, 23	not determined
Fresh Tomatoes	<1, 2, 1	not determined

[a] same brand as can 1.

The average recovery for canned tomatoes was 93 % at the 10 ppb level, n=4, c.v. =9%. However, repeatability and recoveries for the vegetable liquids were poor. Brotons, who reported variability from ca. 30 to 100% with his analysis of vegetable waters for BPA (*9*) made similar observations. The matrix of canned tomatoes is relatively clean when compared to matrices of the vegetable liquids. Even with the shortcomings with vegetable waters, the data show that low parts-per-billion levels of BPA, 4 to 23 ng/mL, migrate from epoxy-based can enamels into vegetable waters and canned tomatoes.

Nonyl phenol determinations. Next, the same type of polymeric SPE cartridge was used for NP determinations. In the first experiment, single side extraction with distilled water of PVC film made with TNPP was investigated. The PVC film was a name brand product purchased at a local grocery store. Films were extracted at 40 °C with a volume to surface area ratio of 45 mL of water to 95.4 cm^2 of the PVC film, and analyzed for NP after 5 and 10 days. These are standard conditions that simulate room temperature storage with no thermal treatment. At the conclusion of the migration period, NP was concentrated on an SPE cartridge, eluted with a few mL's of hexane, and the final extract was analyzed by capillary GC with mass selective detection in the selected ion monitoring mode. The results of the analyses are presented in table IV.

Table IV. Nonyl Phenol Migration from PVC Film into Water

Extraction Time	Migration (ng/mL)
5 days	4.4, 4.5, 4.4, 4.3
10 days	7.2, 7.6, 7.3, 8.5

Recovery studies of NP from water were conducted at the 5 ng/mL level. Recoveries were 88 %, n=4, c.v. =8 %. The data show that after 10 days of storage at 40 ° C (standard conditions that simulate room temperature storage with no thermal treatment), up to 8.5 ng/mL of NP migrates from PVC film into water.

Migrating NP was also measured in some baby juices and aseptically packaged juices. One major brand of baby juices in glass jars with vinyl jar lid gaskets and juices in paperboard cartons with olefin base food contact surfaces (FCS) were analyzed for NP using the same procedure for NP in water. The packages had been found to contain NP residues. The results of the analyses are summarized in table V.

Table V. Nonyl Phenol Migration from Jar Lid Gaskets and Olefin Base Food Contact Surfaces into Fruit Juices

Description	Migration (ng/mL)
Canned apple (control)	<0.2[a]
Apple brand A olefin FCS	2.6
Apple brand B olefin FCS	7.0
Baby apple brand A	19.1, 25.8
Baby pear brand A	26.8, 39.5
Baby white grape brand A	20.1, 23.2

[a] Detection limit = 0.2 ng/mL

Repeatability for NP determinations in the olefin base packaged apple juice brand A at the 2.6 ng/mL level was, n=6, c.v. =17%. Recoveries from the same juice at the 5 ng/mL level were 100 %, c.v. =13%. The data show NP migration into fruit juices from vinyl jar lid gaskets and paperboard cartons with olefin base FCS's (aseptically packaged) ranged from 2.6 to 39.5 ng/mL.

In the final NP experiment, migration into ready-to-feed infant formula from vinyl jar lid gaskets containing NP was investigated. The same extraction procedure was used, but SPE recoveries were < 10%. Recoveries improved with increased dilution of the formula with water. When a 4-fold dilution was used, recovery was only 10%. When dilution was increased 10-fold, recovery increased to 32%. We suspect the NP, being highly lipophilic, has a much greater affinity for the fat in the infant formula than for the cross-linked styrene divinyl benzene solid phase. Table VI summarizes the NP in ready-to-feed infant formula data.

Table VI. Nonyl Phenol Migration from Vinyl Jar Lid Gaskets into Infant Formula

Formula Type	Migration (ng/g)[a]
Ready-to-feed, brand A Canned, milk base (control)	<1[b]
Ready-to-feed, brand A Nursette, milk base	81
Ready-to-feed, brand B Nursette, milk base	77
Ready-to-feed, brand C Nursette, soy base	54

[a] Adjusted for low recoveries.
[b] Detection limit of 1 ng/mL.

Recovery studies were conducted by fortifying the canned ready-to-feed formula with NP. At the 10 ng/mL level, recoveries were 34 %, n=5, c.v. = 14%. Even with the low recoveries, the data are repeatable, and NP migration occurs from vinyl jar lid gaskets into infant formula at levels up to 81 ng/mL.

Phthalates in Infant Formula. In the phthalates in infant formula field assignment, a total of 81 samples were collected at the retail level. Samples were collected from at least 4 different lots for each of three forms; ready-to-feed, powder, and nursettes. At least 4 units were collected from each lot or type. Also, 10 different brands made by 4 different manufacturers sold in the United States were represented. Samples were composited by brand, type, and form with a total of 29 composites analyzed. Table VII summarizes the phthalates in infant formula data.

Table VII. Results of Phthalates in Infant Formula Field Study

Form/# Samples[a]		DBP (ng/g)	DEHP (ng/g)
Brand A milk base w/iron		n.d.[b]	51
Nursette	1 sample		
Brand A milk base		7	n.d.
Powder	2 samples		
Brand A soy base w/iron		11	21
Powder	4 samples		
Brand B milk base w/iron		n.d.	15
Powder	4 samples		
Brand B milk base w/iron		n.d.	<12
Ready to feed	1 sample		
Brand C soy base w/iron		n.d.	30
Powder	4 samples		

[a] Number of samples in composite analyzed.
[b] n.d.= not detected.

Phthalate residues reported in Table VII have been adjusted for background levels found in replicate blanks. The calculated limits of detection were 5 ng/g for DBP and BBP, and 12 ng/g for DEHP. These limits represent quantities above the background levels seen in replicate blanks. Recoveries at the 100 ng/g level for 5 samples representing all manufacturers, brands and forms were 104, 96 and 68 % with c.v.'s of 6, 17 and 15% for DBP, BBP and DEHP respectively. The data show only 6 of the composites were positive for 2 of the phthalates. Di-2-ethylhexyl phthalate was found in 5 of the six at levels from 11 to 51 ng/g. Di-n-butyl phthalate was found in 2 of the composites at 7 and 11 ng/g, and BBP was not detected in any of the composites.

The jar lid gasket of the brand A milk base formula with iron nursette from Table VII was extracted with hexane and the extract analyzed by capillary GC with mass selective detection. No DEHP was detected in the extract at a detection limit of 0.05 µg/mL. When the other containers of infant formula composites that tested positive for phthalates were analyzed, no phthalate residues were detected in any of the containers. These extraction data suggest that the phthalates found in the infant formula were not packaging related. Also, the phthalates measured in the infant formula from the FDA study were orders of magnitude lower than the levels reported in the 1996 MAFF study in Great Britain. The low frequency of occurrence and low levels of phthalates measured in the 1996 FDA infant formula analyses may be because phthalates are ubiquitous in the environment, and they would be expected to be found at least at these low levels and frequencies.

Summary

This work is a continuing process. The data represent efforts completed from 1996 through 1998. Although levels of these chemicals in food packaging can be hundreds of parts-per-million, as in the case of NP in PVC wraps or tens of parts-per-million in the case of BPA in PC bottles, for the most part, only low parts-per-billion levels were found in foods. These low migration levels are achieved under widely different conditions of use; room temperature storage for PVC films, refrigerated temperatures for cheese wrappers, and thermal processing temperatures for infant formula, baby foods, and canned vegetables.

Analysis of food packages in current use have shown a reduction, and in some cases elimination of suspected ED's. Specifically, we have not seen the use of any monomeric plasticizers of any kind in any baby food jar lid gasket tested. Although, we still see NP residues in commercial PVC wrap, the one major household PVC wrap found to contain NP residues in the past no longer contains NP. Two years ago, the jar lid gaskets of 2 major baby food lines contained NP residues, now the jar lid gaskets of only one of the 2 lines contain NP residues.

Phthalate esters are ubiquitous in the environment, but our survey data show that only a few food packages contained phthalate residues and these residues are only in the low parts-per-million range.

Literature Cited

1) Colborn, T., vom Saal, F.S., and Soto, A.M., Environ Health Perspect. **1993**, 101(5), 378-384.

2) Colborn, T, In Our Stolen Future: Are We Threatening Our Fertility, Intelligence, and Survival? A Scientific Detective Story; Dumanoski, D., Myers, J.P., Eds.; Dutton Books, New York, NY, **1996**, 306pp.

3) Gandara, J. , Abuin, s., Losada, P., Lozano, J., J. Chromatog. Sci. **1990**, 31, 450-454.

4) Krishnan, A.V., Stathis, P., Permuth, S.F., Tokes, L., and Feldman, D., Endocrinology, **1993**, 132(6), 2279-2286.

5) Nagel, S.C., vom Saal, F.S., Thayer, K.A., Dhar, M.G., Boechler, M., and Welshons, W.V., Environ Health Perspect, **1997**, 105(1), 70-76.

6) Soto, A.M., Justicia, H., Wray, J.W., and Sonnenschein, C., Environ Health Perspect, **1991**, 92, 167-173.

7) Jobling. S, Sumpter, J.P., Aquatic Toxicol, **1993**, 27, 361-372.

8) Purdom, C.E., Hardiman, P.A., Bye, V.J., Eno, N.C., Tyler, C.R., and Sumpter, J.P., Chem Ecol, **1994**, 8, 275-285.

9) Brontons, J.A., Olea-Serrano, M.F., Villalobos, M., Padraza, V., and Olea, N., Environ. Health Perspect, **1995**, 103(6), 608-612.

10) Jobling, S., Sheahan, D., Osborne, J.A., Mathiessen, P., and Sumpter, J.P., Environ Toxicol Chem, **1996**, 15, 194-202.

11) Sharpe, R.M., Fisher, J.S., Millar, M.M., Jobling, S. and Sumpter, J.P., Environ Health Perspect, **1995**. 103(12),1136-1143.

12) Mylchreest, E, Cattley, R.C., and Foster, P.M.K., Toxicological Sciences, **1998**, 43, 47-60.
13) *Code of Federal Regulations* (1998) Title 21, 175-178, and 181, U.S. Government Printing Office, Washington, DC.
14) Biles, J.E., McNeal, T.P., Begley, T.H., and Hollifield, H.C., J. Agric. Food Chem., **1997**, 45, 3541-3544.
15) Ministry of Agriculture, Fisheries and Food of Great Britain, Food Safety Directorate, **March 1996**, Phthalates in Infant Formulae, Food Surveillance Information Sheet 83.
16) *Consumers Reports*; editors, Eds.; Hormone mimics *They're in our food; should we worry*?, Consumers Union: New York, NY, June 1998, 52-55.
17) Biles, J.E., McNeal, T.P., Begley, T.H., J Agric. Food Chem. **1997**, 45, 4697-4700.
18) *Pesticide Analytical Manual*; McMahon, B.M., Wagner, R.F., Eds.;Volume 1: Multiresidue Methods, 3rd Edition; 1994, United States Department of Human Services, Public Health Service, Food and Drug Administration: Washington, DC, **1994**. Also available via the Internet at http://vm.cfsan.fda.gov/~ frf/pami1.html.)

Chapter 5

Environmental Exposures to Agrochemicals in the Sierra Nevada Mountain Range

James LeNoir[1], Linda Aston[2], Seema Data[3], Gary Fellers[4], Laura McConnell[3], and James Sieber[5]

[1]**Center for Environmental Sciences and Engineering, Mail Stop 199, University of Nevada, Reno, NV 89557**
[2]**Pacific Toxicology Laboratories, 1545 Pontius Avenue, Los Angeles, CA 90025**
[3]**ECL, Agricultural Research Service, U.S. Department of Agriculture, 10300 Baltimore Boulevard, Beltsville, MD 20705**
[4]**Biological Resources, U.S. Geological Survey, Point Reyes Seashore, Point Reyes, CA 94956**
[5]**Western Regional Research Center, Agricultural Research Service, U.S. Department of Agriculture, Albany, CA 94710–1105**

The release of pesticides into the environment may impact human and environmental health. Despite the need for environmental exposure data, few studies quantify exposures in urban areas and even fewer determine exposures to wildlife in remote areas. Although it is expected that concentrations in remote regions will be low, recent studies suggest that even low concentrations may have deleterious effects on wildlife. Many pesticides are known to interfere with the endocrine systems of humans and wildlife, adversely affecting growth, development, and behavior. This chapter reviews the fate and transport of pesticides applied in the Central Valley of California and quantifies their subsequent deposition into the relatively pristine Sierra Nevada Mountain Range.

California presents a unique setting for examining the impact of pesticides on non-target sites. The region's Mediterranean climate supports a diverse year-round agricultural industry. The diversity of crops leads to an intensive pesticide management schedule; more pesticides are applied per acre in California's Central Valley than anywhere else in the United States. The Central Valley is enclosed to the west by coastal mountains and to the east by the Sierra Nevada Mountain range. Prevailing weather patterns flow east from the Central Valley and up the slopes of the mountains carrying dust, soot, and vapors and redepositing them in the ecosystems of the Sierra Nevada mountains (Figure 1). The deposition of pollutants in the Sierra Nevada is of concern because of the potential impacts the pollutants may have on wildlife and plants.

During the last few decades native frog populations throughout California have declined precipitously and efforts to determine the cause of this decline have been inconclusive. Many hypotheses have been advanced to explain the alarming rate of frog decline, such as an increase of UV-B radiation (*1*), introduction of non-

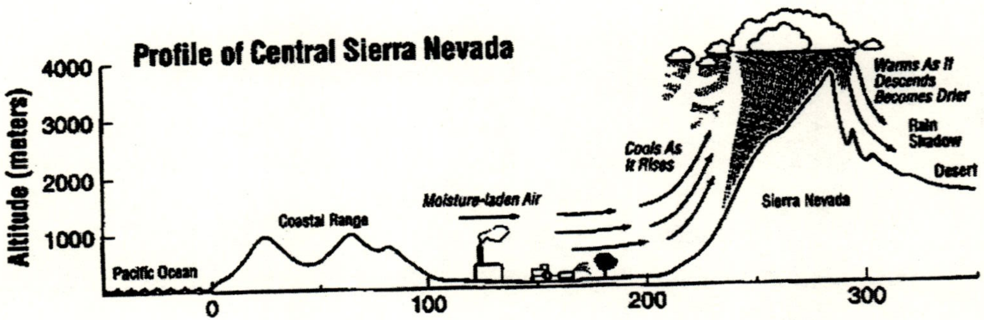

Figure 1. Cross section of California, illustrating transport of pollutants to Sierra Nevada mountain range. (Reproduced with permission from reference 17. Copyright 1993.)

native species (*2*), pH effects (*3*), and pesticides (*4*). Various studies have tested each hypothesis, although most studies conclude no single hypothesis can explain the decline of amphibian populations. Dwindling frog populations are of concern not only for the loss of biodiversity but also because they are perceived as sensitive indicators of future environmental repercussions. Some researchers speculate that pesticides may be a contributing stress due to their toxicity but offer little or no supporting data.

This chapter provides an overview of what is known regarding atmospheric transport of pesticides from California's Central Valley to the Sequoia National Park on the southwestern slope of the Sierra Nevada Mountain range in Tulare county (Figure 2). The processes involved are volatilization, movement from the target source and subsequent deposition. More comprehensive reviews of the environmental fate of pesticides are available (*5,6*). Here, we discuss the unique features of pesticide release and subsequent transport to the nearby Sierra Nevada mountain range and assess the potential impact on declining frog populations.

Pesticide Use in the Central Valley

All California agricultural communities could serve as emission sources of pesticides entering the Sequoia National Park. The Sequoia National Park is located in the southwest corner of Tulare county (Figure 2.) The park is downwind of four intense agricultural counties: Madera, Fresno, Kings, and Tulare. These four counties are likely to serve as the greatest sources of pesticides entering the park. The total amount of active ingredient applied annually in these four counties is listed in Table 1. The total mass applied is indicative of the source strength. Sampling sites were established at different elevations within the Kaweah river basin of the park (Figure 3) to determine the efficiency of transport as distance and elevation from the source increased.

A wide variety of pesticides are applied in the Central Valley. Our research has focused on the current-use pesticides and a few historic pesticides such as p,p'-DDT illustrated in Figure 4. One of the most abundantly applied classes of pesticides in the Central Valley is the organophosphates. Organophosphates are broad-spectrum insecticides used on crops prior to harvest, but uses also include homes, gardens, turf, ornamentals, and forestry. Beginning in the 1960's, organophosphates largely replaced the persistent organochlorine compounds such as DDT. The organophosphates determined in our studies included diazinon, chlorpyrifos, methidathion, and malathion. Endosulfan is one of the few organochlorine insecticides still approved for use. It is sold as an isomeric mixture of 70:30 α-endosulfan and β-endosulfan. Endosulfan is a broad-spectrum insecticide used on fruits, vegetables and cotton. Chlorothalonil is a nonsystemic foliar fungicide applied to crops to protect them from mildews, leaf spot and blights. Chlorothalonil and endosulfan are less persistent organochlorine pesticides than DDT and other banned organochlorine chemicals, but generally more persistent than organophosphate pesticides. Trifluralin is an herbicide of the trinitroaniline family; it is a pre-emergent broadleaf herbicide that is plowed into the soil prior to planting.

Figure 2. Map of California illustrating the four agricultural counties studied.

Table 1. Pesticide use data and environmentally significant properties of target pesticides.

Pesticide	Annual Application (Kg) (*53*)	Emission Soil (*44*)	Rate(g/m^2*sec) Plants (*44*)	Vapor Pressure (mPa)	Solubility (mg/L)	Log K_{ow} (*52*)
Chlorpyrifos	1,081,003	$2.56*10^{-8}$	$2.56*10^{-8}$	2.5 (*44*)	1.18 (*52*)	5.0
Diazinon	678,050	$7.47*10^{-8}$	$7.47*10^{-9}$	16 (*47*)	60 (*52*)	3.30
Malathion	325,551	$5.70*10^{-11}$	$5.70*10^{-11}$	1.1 (*45*)	145 (*23*)	2.7
Methidathion	205,983	$2.11*10^{-11}$	$2.11*10^{-11}$	0.45 (*46*)	240 (*35*)	2.2 (*35*)
α-Endosulfan	166,367	$5.76*10^{-8}$	$1.26*10^{-7}$	6.2 (*23*)	3.7 (*23*)	3.13
β-Endosulfan				3.2(23)	21(*23*)	3.62
Trifluralin	638,221	$1.67*10^{-6}$	$1.67*10^{-6}$	15 (*46*)	0.32(*23*)	5.07
Chlorothalonil	495,750	$2.25*10^{-8}$	$2.25*10^{-8}$	0.3 (*46*)	0.056	2.88
DDE	DDT Banned 1971	ND	ND	0.30 (*35*)	0.04(*35*)	5.7(*35*)

Seiber (*23*)
Schmidt (*35*)
Suntio (*44*)
Kim (*45*)
Woodrow (*46*)
Taylor (*47*)
Herner (*52*)
State of California Pesticide Use Report (*53*)

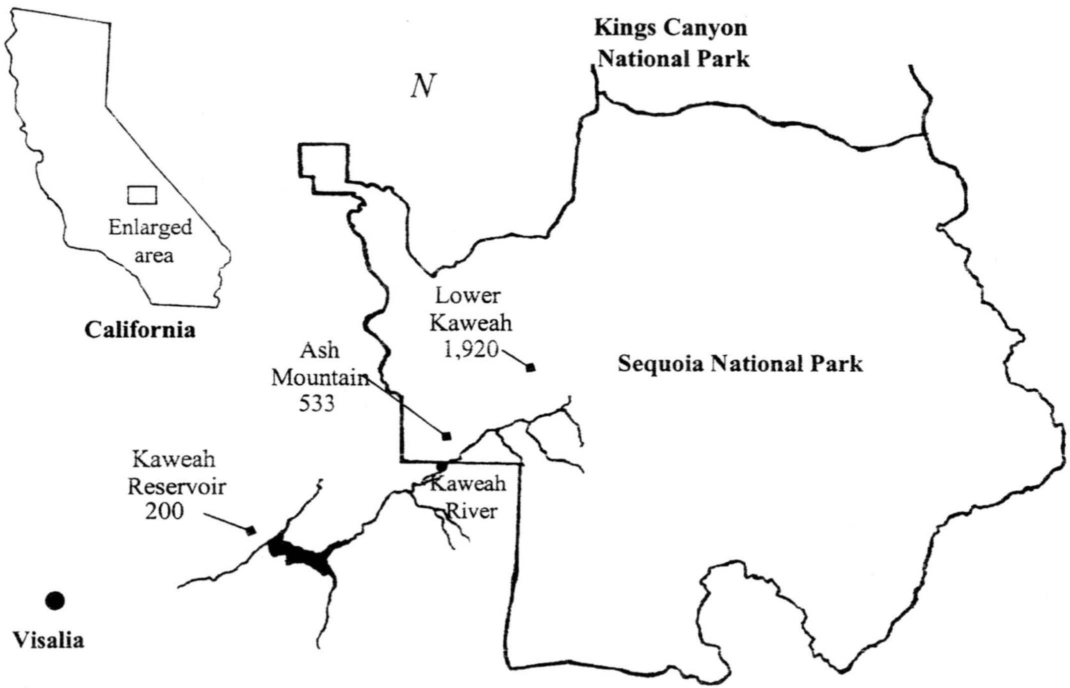

Figure 3. Map of Sequoia National Forest, California, USA. Locations of sampling are designated with elevation in meters.

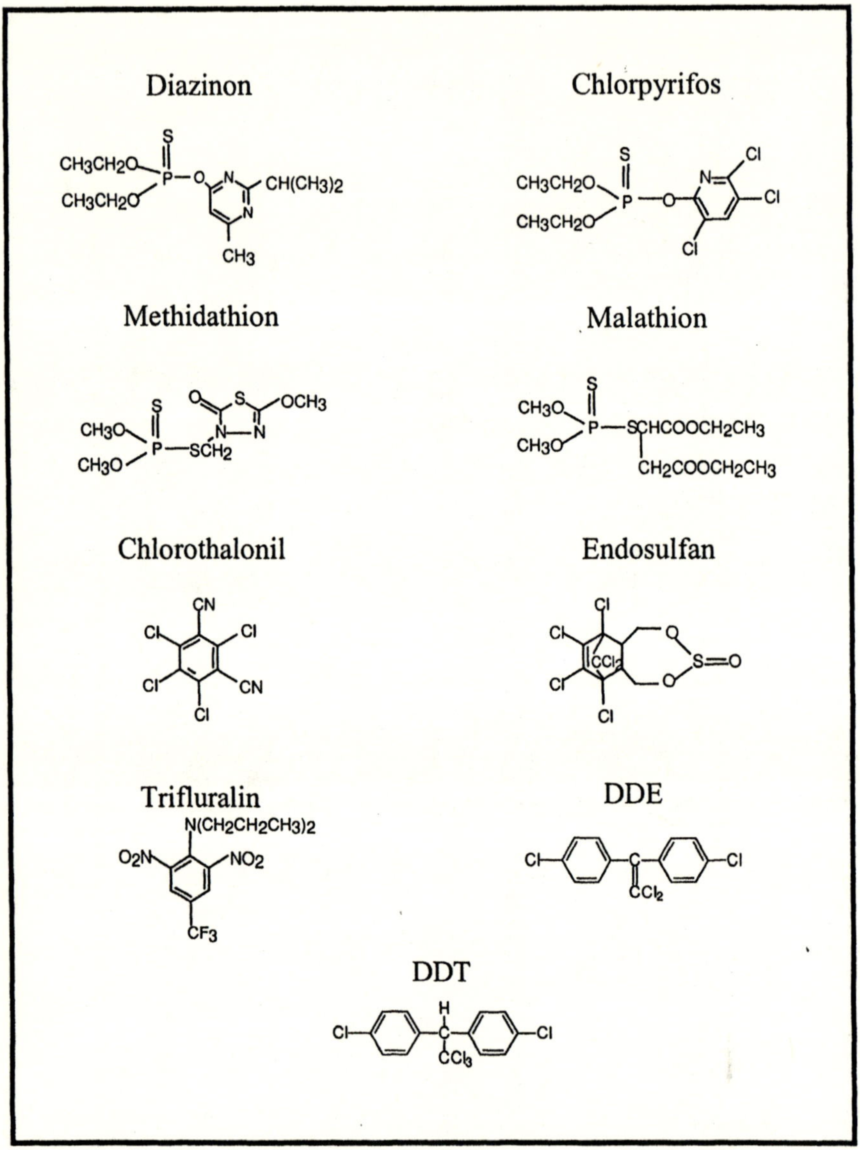

Figure 4. Chemical structures of pesticides investigated in this chapter.

Some widely used chemicals became ubiquitous contaminants in the environment. DDT is such a compound. DDT was widely used for disease vector control and crop protection in the U.S. and worldwide beginning in the early 1940s. It was later found to be environmentally persistent and induce adverse ecotoxicological effects. The best known effect is eggshell thinning in birds (*7*). It was consequently banned from commercial use in the U.S. by 1973, but it is still employed in other parts of the world. The predominant form of DDT in the environment is its breakdown product DDE. The presence of DDE in the biota of the Sierra Nevada to this day is indicative of its environmental persistence.

Transport and Transformation

In order for the pesticides to be present at non-target sites, they need to be transported from the site of application. Pesticide volatilization begins at the point of application. Pesticides enter the air mass through spray drift, volatilization and wind blown soil erosion. The potential for volatilization is dependent upon the method of application, the physico-chemical properties of the pesticide and environmental factors such as temperature wind and soil moisture content. Environmentally significant physical properties are listed in Table 1.

Pesticides can be introduced into an airmass as vapors or bound to particulate matter. Once in the atmosphere, pesticide vapors can remain free or adsorbed to particulate matter. As particulates, pesticides may be removed by gravitational settling or turbulent transfer by impaction upon a surface. The deposition rate of particulate matter is size dependent. Large particles having an aerodynamic diameter > 2 μm rapidly settle due to gravitational forces while smaller particles settle more slowly and thus are transported greater distances (*8*).

Physical properties such as vapor pressure, water and lipid solubilities influence movement and the distribution of pesticides in the environment. Chemicals that have high vapor pressures volatilize more effectively than compounds with lower vapor pressures. The surface to which pesticides are applied also influences their fate; the volatilization of some pesticides may increase by an order of magnitude or more when applied to vegetation rather than to soil. As much as 90% of some pesticides may volatilize from the application site in as little as three days (*9*).

Equilibrium partitioning of a chemical between octanol and water, described by the octanol-water partition coefficient (K_{ow}) is indicative of the chemical's ability to bioaccumulate in the food chain. Pesticides that have a high K_{ow}, and are environmentally stable, such as DDT, tend to accumulate in the food chain (*10*). The Henry's law constant reflectss the equilibrium partitioning of a compound between air and water. In the natural environment, equilibrium of a pesticide between two compartments is not always achieved; however, partition coefficients are still used to explain which environmental compartment is favored. Given the large quantities of pesticides applied, and the potential for movement from the target source, there is enhanced interest in knowing the environmental fate and non-target effects of these pesticides.

During the course of pesticide movement from air-water-soil matrices, chemical and biological processes transform the chemical into secondary products.

The extent and potential for chemical transformations is dependent upon the residence time in a given matrix and the inherent reactivity of the chemical. The fate of pesticides in the atmosphere includes removal by degradation. Photooxidations, both those resulting from direct photolysis and chemical oxidation, are the most important reaction pathways. These degradative pathways can also include reactions with water, hydroxyl radical, ozone and nitrate chemistry. The importance of each reaction pathway will depend on the light absorbing and physico-chemical properties of the chemical of interest.

Some atmospheric reactions lead to pesticide activation, that is, a change in chemical structure that creates byproducts of equal or greater toxicity and persistence than the parent. A change in the chemical structure may cause the pesticide to become environmentally persistent, to increase in lipophilicity or to become more mobile, consequently posing a greater threat to non-target organisms (*11*). Of the pesticides under consideration the organophosphate pesticides undergo such environmental transformations.

Organophosphate pesticides are activated in the environment from the parent thion form to the oxon form. The oxon is formed when oxygen is substituted for the sulfur doubly bonded to the phosphorous atom. The mechanism for this transformation in the air is not understood but is suspected to involve reaction with atmospheric oxidants and reactive aerosol surfaces (*12,13*). The substitution of oxygen for sulfur increases the toxicity, causing greater acetylcholinesterase inhibition in nerve tissue. The increased toxicity for the insecticde ethyl parathion was realized in declining populations of red-tailed hawk (*Buteo jamaicensis*) surrounding orchards in the Central Valley (*14*). In 1990 ethyl parathion was banned in California due to its human toxicity and potential threat to farm workers. When parathion was banned, red-tail hawk populations surrounding the orchards recovered.

Air Deposition. The transport of pollutants from the Central Valley to the Sierra Nevada Mountain range is dependent upon the season. During the winter months, the transport of pollutants to the Sierras is largely prevented by inversions over the Central Valley (*15*), and reduced temperatures in the valley and mountains. Transport of pesticides is attenuated during the winter months, however, when storm fronts pass through the Central Valley, the inversion is broken and transport is increased. Air and rain samples collected during the winter of 1990-1991 illustrate the minimal transport that occurs during winter months (Table 2.) (*16,17*). Air concentrations within a citrus orchard 17-Km outside the entrance to the Sequoia National Park are two to three orders of magnitude higher than the air concentrations reaching the park entrance. Chlorpyrifos averaged 2 ng/m^3 near treated orchards in the month of February whereas 8 pg/m^3 was the average air concentration at the park entrance during the same month. Similar trends were seen for diazinon and diazinon oxon.

During the summer months, however, air quality in the Southern Sierra Nevada Mountain range is significantly degraded (*18*). The reason for the reduction in air quality is the increased transport of pollutants; stable inversions over the Central Valley are not as common during summer months, therefore

Table 2. Concentrations of organophosphate compounds in winter air and rain at two elevations (*15*).

		Air Concentrations (pg/m^3)			Rain Concentrations (ng/L)		
Month	Elevation (m)	Chlorpyrifos	Diazinon	Diazinon oxon	Chlorpyrifos	Diazinon	Diazinon oxon
Lindcove, CA							
December	114	1,300	42	17	18	78	14
January	114	1,500	206	64	180	2,000	320
February	114	2,300	1,221	483	<LOQ	6,100	2300
Ash Mountain, CA							
December	533	9	4	5	3	4	<LOQ
January	533	7	3	4	<LOQ	<LOQ	<LOQ
February	533	8	6	5	13	<LOQ	5

atmospheric pollutants are not restricted to the Central Valley. In summer months, concentrations of pesticides in the air mass are higher due to increased agricultural production and higher temperatures which increase volatilization of pesticides from sources. The summer air concentrations of chlorpyrifos and methidathion for the summer of 1994 are listed in Table 3. We found that the concentrations of chlorpyrifos in the citrus orchard were one to two orders of magnitude higher than we had observed in the 1991 winter air listed in Table 2. The highest summer air concentration in the orchard was 40,000 pg/m^3, whereas during winter the highest average value was 2,300 pg/m^3 for the month of February, approximately a 20-fold difference. The increased concentrations at the park entrance are not representative of the whole park. Air concentrations at the Lower Kaweah site were even lower than those found at Ash Mountain. Methidathion concentrations were lower at all sites because it is not used as widely as chlorpyrifos. The significant drop in methidathion concentrations at downwind elevations can be explained by the fact that methidathion is more effectively removed from the atmosphere by oxidative reactions. Methidathion has an atmospheric half-life of 0.86 hours whereas chlorpyrifos has a half-life of 1.4 hours (*19*).

Vegetation. As the air mass moves into the Sierras, airborne pollutants can interact with vegetation. The influence that terrestrial vegetation may have on the fate of vapor phase toxics is key to understanding the effects of these compounds on terrestrial ecosystems. The extent of interaction between vapor phase pesticides and plant foliage is largely unknown. Pesticides may deposit onto foliage while bound to dust particles. Pesticides vapors may adsorb directly to the surface and enter the cuticle, or penetrate further to the interior of the leaf. The accumulation of toxic compounds from the atmosphere into plant tissues has received attention, largely because it is recognized that consumption of plants may be a route of exposure for humans and animals.

In our studies, Ponderosa pine (*Pinus ponderosa*) needles were collected at 112, 533, and 1,920-m sites during the summer months to determine the relative importance of adsorption of pesticides onto vegetation. During the summer of 1994 concentrations in the needles were found to follow the same trend as the air concentration data (Figure 5 and 6). The site within the agricultural field has relatively high levels of parent organophosphates and their oxon degradation products. At higher elevations, however, pine needles contained low to immeasurable concentrations of organophosphate residues. The predominant form of the pesticides found to survive to higher elevations via the air were the oxon homologs of these organophosphate pesticides (*20,21*). These results suggest that foliar absorption is minimal but should be considered when assessing the downwind fate of pesticides.

Wet Deposition. Rain and fog concentrate suspended particulate matter and vapors as they travel through the atmosphere and subsequently deposit the contaminants on the surfaces they impact, a phenomenon known as washout (*22, 23*). Single event rain sampling at the 533-m and 1920-m sites within the Sequoia National Park indicate that rain acts as a concentrating agent (Table 4)(*24*). The

Table 3. Summertime air concentrations of chlorpyrifos and methidathion (ng/m^3)(*21*).

		Air Concentrations (ng/m^3)			
Month	Elevation (m)	Chlorpyrifos	Chlorpyrifos Oxon	Methidathion	Methidathion Oxon
Lindcove, CA					
June	114	40	32.0	10.2	8.35
July	114	24.6	17.1	4.3	4.66
August	114	9.9	7.1	0.98	1.45
Ash Mountain, CA					
June	533	0.490	2.18	0.230	0.624
July	533	0.120	1.70	N.Q.	0.210
August	533	<LOQ	0.37	N.Q.	N.Q.
Lower Kaweah, CA					
June	1,920	0.219	0.570	N.Q.	0.210
July	1,920	0.120	0.316	N.Q.	<LOQ
August	1,920	<LOQ	0.128	N.Q.	N.Q.

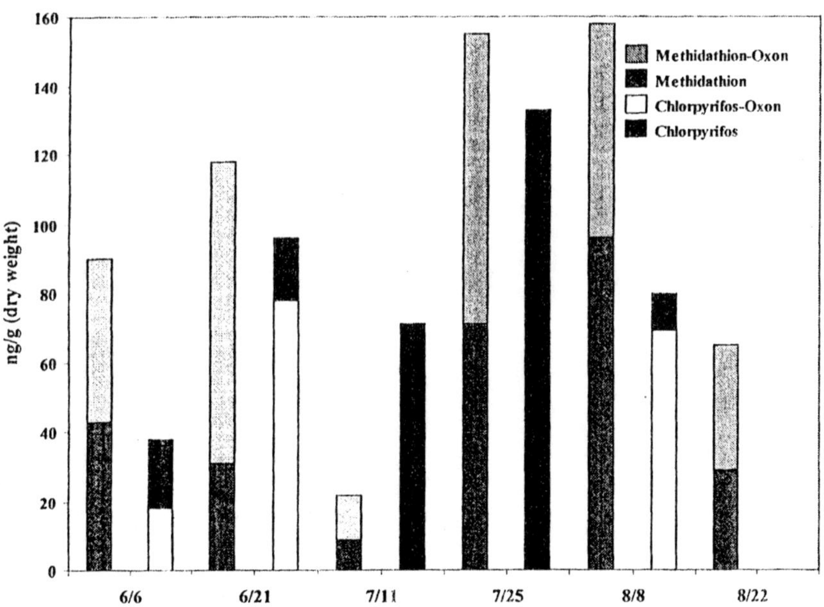

Figure 5. Organophosphate residues in pine needles sampled agricultural field (112 m.) (*20*).

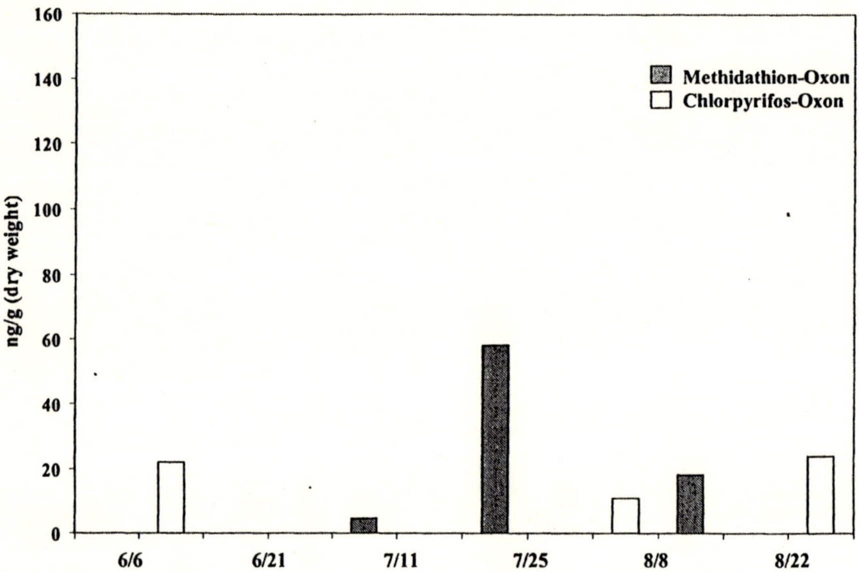

Figure 6. Organophosphate residues in pine needles sampled at Ash Moutain (533 m.) (*20*).

Table 4. Pesticide Concentrations in Rain and Snow from the Sierra Nevada Mountains, (ng/L) (*24*). Reprinted with Permission from (Environmental Expsoures to Agrochemicals in the Sierra Nevada Mountain Range, McConnel, et al.) Copyright Society of Environmental Toxicology and Chemistry (SETAC), Pensacola, FL, 1997.

533-m elevation	Trifluralin	Chlorothalonil	Chlorpyrifos	Diazinon	Malathion	Endo-I	Endo-II
12/14/95	<0.098	0.73	1.3	4.4	7.7	1.6	0.64
1/18/96	<0.098	0.97	2.3	19	3.6	<0.04	0.23
1/24/96	<0.098	<0.40	4.4	7.0	<0.05	0.86	0.27
2/1/96	<0.52	3.3	2.1	<0.31	<0.25	<0.20	<0.25
2/26/96	1.2	85	4.4	6.8	24	6.5	1.4
3/30/96	<0.87	31	1.3	<0.51	<0.41	<0.31	<0.42
4/2/96	<0.35	13	3.1	<0.21	<0.17	<0.13	<0.17
1920-m elevation							
1/18/96	<0.16	1.1	4.9	14	6.0	0.97	<0.08
2/6/96	<0.098	<0.57	1.1	<0.06	<0.05	0.93	0.19
2/20/96	0.8	2.74	2.2	3.7	3.7	2.8	0.26
3/5/96	0.52	2.29	2.5	6.5	<0.05	1.0	<0.01
4/2/96	2.4	12.96	13	4.5	6.0	3.0	0.46

efficiency of washout is related to the chemicals' Henry's law constant. Compounds with a low vapor pressure and high water solubility (i.e. low Henry's constant) favor washout.

Biota. We have detected pesticides in air, wet deposition, and vegetation in the Sierras; the residue in each of these environmental compartments has the potential to collect in surface waters. Both vapor-phase and particulate-phase pesticides can be scavenged from the atmosphere by rain, snow, and fog and be deposited into surface water. Wet deposition (rain and snow) can also wash plant and soil surfaces and transport residues to surface waters. A pesticide in the vapor phase can undergo some direct air-water partitioning based on its Henry's Law constant. Consequently, surface waters can serve as potential sites of pesticide concentration.

Biologically and chemically stable pollutants have the potential to accumulate in the food chain. The result of this accumulation is that the concentration in an organism is greater than that in its environment. For aquatic organisms, this can occur by two routes. One route is through the process called bioconcentration, which is direct partitioning from water into the organism. The second, biomagnification, is accumulation through the organism's diet. The degree to which these processes occur is directly related to the hydrophobicity of a chemical, the lipid content of the organism, and its life habits.

To assess aquatic exposures, we examined pesticide residues in Pacific tree frog (*Hyla regilla*) tadpoles and in brook trout within the Sequoia National Park, during the summer of 1996 (*25*). Pacific tree frog tadpoles were collected from field enclosures at Sycamore Creek, 488-m elevation in the Park. The tree frog was used as a surrogate for the endangered frog species, Mountain Yellow-legged frog (*Rana muscosa*) and Red-legged frog (*Rana aurora draytonii*).

Analyses for pesticides in Pacific tree frog tadpoles revealed the presence of two widely used agricultural pesticides, the organophosphorus pesticide chlorpyrifos and the fungicide chlorothalonil (Table 5). Chlorothalonil was present at slightly higher levels than chlorpyrifos. Application amounts in the San Joaquin Valley area were ca. 350,000 kg and 100,000 kg respectively for chlorpyrifos and chlorothalonil. Given the high use, the presence of these pesticides in tadpole tissues is not surprising and is indicative of these organism's to concentrate aqueous concentrations of these hydrophobic substances by physical partitioning.

In addition, the tadpoles were examined for the historic-use organochlorine DDT and its breakdown product DDE. Only residues of DDE were observed. The concentrations of p,p'-DDE were significantly higher than the current-use pesticide residues. These concentrations of DDE in trout and tadpoles can be explained by past heavy use of DDT in the Central Valley and earlier uses in forest management.

Three brook trout were collected from the Kaweah River at 700-m elevation using hook and line. The trout were not examined for current-use pesticides. Levels of p,p'-DDE in Kaweah River trout ranged from 40.1 to 65.7 ppb and averaged 55.3 ppb. The concentration of DDE in brook trout was an order of magnitude higher than in tadpoles. The higher concentrations of DDE in trout can be explained by the greater exposure of the trout. The tadpoles had less time to accumulate pesticides due to their short exposure before collection whereas the brook trout, being more mature had more time to accumulate organics.

Based on these results we speculate that the variability may be due to different levels in surface waters, or it may indicate differences in metabolism, diet, and habitat. Because the concentrations of lipophilic compounds such as DDE may be higher in shallow waters than in deeper waters, egg masses and tadpoles may be receiving greater exposure than fish due to their respective habitats. *Hyla regilla* eggs are laid on emergent vegetation in shallow water and the tadpoles bask and feed in shallow waters (*26*).

Environmental Impact

Acute Effects. There are indications that the wildlife of the Sierra Nevada Mountain range are under stress. The Mountain Yellow-legged frog and Red-legged frog once abundant in the Sierra Nevada Mountain range are noticeably declining. The Red-legged frog historically inhabited a range from the coast to 1800-m elevation in the Sierras. The Red-legged frog is no longer present in 99% of its historic range within the Sierras (*27*). In 1996 the U.S. Fish and Wildlife Service listed the Red-legged frog as a threatened species. The Mountain Yellow-legged frog was once common at and above 1500-m elevation, but is now absent at many sites where it was once common (*28-31*). Attempts to reintroduce the Mountain Yellow-legged frog at higher elevations have failed (*32*). This suggests that whatever was causing the declines is still present.

Pesticides are suspected as causative agents since fish and frogs have been found to accumulate organic pollutants emanating from urban and agricultural areas. Amphibians are important indicators of environmental contamination because they are present in both aquatic and terrestrial environments. The scarcity of acute and chronic amphibian toxicity data limit our ability to assess a toxic impact or risk. The toxicity data presented in Table 6 is a summary of published values for frogs in the genus *Rana*, (the same genus to which the declining frogs of the Sequoia National Park belong). The values listed vary due to interspecies and lifestage sensitivities as well as testing differences but they yield a range of values that can be used to assess the impact of pesticides on frog populations.

Of the organophosphorous compounds that we have studied, only data for malathion and diazinon toxicity were available. The LC_{50} concentration for malathion's active ingredient is 150 mg/L. This concentration is much higher than the highest malathion concentration of 81 ng/L we found in surface waters. The magnitude of malathion toxicity to ranid frogs is also consistent with toxicity values of other organophosphorus pesticides on related ranid frogs (*33,34*). The LC_{50} concentration for diazinon is significantly lower than that of malathion. The concentrations of diazinon do not reach the acute toxicity level within the forest; however, there may be brief toxic exposures when rain and fog introduce additional concentrations. Diazinon is applied in great quantities in the Central Valley, particularly in the winter months of January and February, when it is applied as a dormant orchard spray. Chlorpyrifos was found in all environmental

Table 5. Concentration of chlorinated pesticides (wet wt.) in brook trout and tree frogs from the Kaweah River Basin, summer 1996. (*25*)

	Collection Date	Sample	Mass (g)	p,p'DDE (ng/g)	Chlorothalonil (pg/g)	Chlorpyrifos (pg/g)
Pacific Tree Frog	6/22/96	31 Tadpoles	4.5	9.6	48	10
Pacific Tree Frog	7/9/96	44 Tadpoles	4.6	9.3	33	17
Brook Trout	7/10/96	Ave. 3 fish	134.3	55.3	NA	NA

Table 6. Toxicity values for various Ranid frogs.

Species	Stage	Chemical	Duratin	Effect	Conc. (μg/L)
R. pipiens (*42*)	Adult	Malathion	48 hour	LC50	150,000
			15 day	83% mortality	175,000
R. tigrina (*42*)	Embryo	Malathion (formulation)	24 hour	100% mortality	40,000
R. calmitans(*51*)	Tadpole	Diazinon	96 hour	LC50	50
			16 day		5
R. clamitans(*51*)	Tadpole	Diazinon (formulation)	96 hour	LC50	25
			16 day		2.8
R. tigrina (*42*)	Tadpole	Endosulfan	96 hour	LC50	1.8
R. sylvatica (*43*)	Tadpole	Endosulfan	96 hour	Reduced avoidance response	32
R. clamitans (*51*)	Tadpole	Endosulfan (formulation)	96 hour	LC50	>11,750 >47,000
			13 day		15
			16 day		15
R. pipiens (*42*)	Tadpole	DDT	12 day	30% mortality	5
			8 day	60% mortality	25
R. temporaria (*42*)	Tadpole	DDT	48 hour	94% mortality	500
R. sylvatica (*42*)	Tadpole	DDT	96 hour	16.51 μg/g in liver at 15°C	1
			7 day	9.43 μg/g in liver at 15°C	1
			96 hour	16.73 μg/g in liver at 15°C	3
			7 day	9.71 μg/g in liver at 15°C	3
			10 day	18.6-23.7 μg/g in fat at 15°C	1
			7 day	26.7-81.7 μg/g in fat at 15°C	3

compartments including the biota; however, toxicity data has not been reported for chlorpyrifos in frogs. It is difficult to predict the toxicity of chlorpyrifos since there is such disparity in the toxic values between diazinon and malathion. Testing is required to determine the toxicity of chlorpyrifos.

Organochlorine chemicals are of particular concern because of their toxicity, persistence and magnification in amphibian food webs. Endosulfan is one of the few historic organochlorine pesticides still used today. It is employed for crop protection because it is not as persistent in the environment as DDT and other members of the organochlorine family. The physical properties of endosulfan are currently under review because only recently have the true structures for the α and β isomers been elucidated (*35*). The hydrolysis half-life of the α-isomer is reported to be 218 hours and 187 hours for the β-isomer at 25 °C and pH 7 (*36*). The concentrations at which endosulfan is toxic to frogs is several orders of magnitude lower than that of organophosphates. At a concentration of 32 µg/L endosulfan reduces the avoidance response of *R. sylvatica*. The 96-hour LC_{50} for endosulfan to *R. tigrina* is 1.8 µg/L. The differences in toxicity between the two frogs suggest an interspecies sensitivity to endosulfan. Toxicity values for DDE are not available for frogs, but an estimate of their impact can be determined on DDT toxicity data. The toxic concentrations for DDT are low, however, its present day effects cannot be ignored.

We conclude that the concentrations of pesticides in the Sierra Nevadas are too low to be inducing acute effects. If the pesticide concentrations reported in this chapter are having a deleterious impact it would have to be chronic.

Sublethal Effects. Growing evidence suggests that many pesticides disrupt the endocrine systems of humans and wildlife (*37-39*). The endocrine disrupting effects of historic-use chemicals such as DDT have been well documented on some species. Historic-use chemicals have manifested reproductive failures in seals in the Netherlands (*38*), the feminization of gulls in southern California (*38*) and abnormal male development in alligators in Florida (*39*); but the effect of current-use pesticides is not as well understood. Of the pesticides reviewed in this chapter, DDT, DDE, chlorothalonil, chlorpyrifos, endosulfan, malathion and trifluralin are suspected endocrine disruptors (*40*). The environmental concentrations reported in this chapter are well below classic toxicity tests where the toxic index is normally death; but classic toxicity tests evaluate one compound at a time, while wildlife is simultaneously exposed to multiple chemicals. The exposure to multiple chemicals allows for additive and synergistic effects. Laboratory studies have shown that when Coturnix quail (*Coturnix coturnix*) were given oral doses of 5 mg/kg DDE and 2.5 mg/kg parathion, parathion toxicity increased in quail by nine-fold (*41*), indicating that pesticides can act synergistically. These concentrations are environmentally relevant but still much higher than those found in the Sequoia National Park. Endocrine disrupting effects of these pesticides cannot be estimated because it is not known at what concentrations, or by what mechanisms these pesticides induce endocrine disrupting effects.

Current-use pesticides may introduce subacute effects that jeopardize an organisms' fitness. A reduction in fitness may be broadly defined as an increase in development time, a reduction in response, a decrease in reproduction or immunosuppresion. To understand the risk these pesticides may introduce, toxicity tests need to be designed which test these environmentally significant endpoints. Species decline, such as that seen in frogs is difficult to study from a cause-effect perspective because several pesticides may interact to have an additive or synergistic effect. The effect of pesticides may occur over several years or several generations rather than immediately. Alternatively, the primary effect may be upon some organism that serves as part of the food chain for the species of interest, rather than the species itself.

Conclusions

Anthropogenic pollution has been implicated as a stress to wildlife. It is tempting to hypothesize that chemicals released from agriculture are causing frog declines in remote areas, but it has yet to be proven. There is sufficient evidence, however, that current-use and historic-use pesticides used in California's Central Valley are present in areas of known amphibian declines in the Sierra Nevada Mountain range. This summary verifies the transport of current-use pesticides applied in California's Central Valley to the Sierra Nevada Mountains and quantifies exposure concentrations in various environmental compartments. Pesticide concentrations in air, wet deposition, and vegetation follow a general pattern of decreasing concentrations with increasing distance and elevation from the source of application. The concentrations of pesticides presented in this chapter are not likely to be causing acute toxicity, but assessing their chronic effect is difficult. Animals in the wild are simultaneously exposed to multiple chemicals through various routes that could contribute additively and act synergistically to contribute to chronic effects. Furthermore, many of the chemicals studied exhibit endocrine disrupting effects. It is understood that endocrine disruptors can act effectively at low concentrations but the threshold concentrations at which their effects are manifested are unknown. The exposure concentrations reported in this chapter should help bridge the gap between field and laboratory tests.

Further research will permit definitive conclusions about the impact of anthropogenic pollutants on wildlife. The presence of these compounds, however, suggests that other pollutants are probably being transported as well and that all chemicals must be considered to assess the complete toxic burden. Clearly more sampling and analysis must be conducted in conjunction with population and toxicity studies to elucidate the impact of pollutants on wildlife.

Literature Cited

1. Hayes MP, Jennings MR. 1986. *J Herpetol* 20:490-509.
2. Licht LE, Grant KP. 1997. *Amer Zool* 37:137-145.
3. Hall RJ, Henry PF. 1992. *Herpetological Journal* 2:65-71.
4. Sadinski WJ, Dunson WA. 1992. *J Herpetol* 26:413-422.
Chemicals in Air, Water, and Soil. John Wiley & Sons.

6. Spencer WF, Farmer WJ, Jury WA. 1982. Review: Behavior of organic chemicals at soil, air, and water interfaces as related to predicting the transport and volatilization of organic pollutants. *Environ Toxicol Chem* 1:17-26.
7. Blus LJ, Gish CD, Belisle AA, Prouty RM. 1979. *Nature* (Lond) 235:376-377.
8. Bidleman TF. 1988. *Environ Sci Technol* 22:361-367.
9. Taylor AW 1978. Postapplication volatilization of pesticides under field conditions. *J Air Poll Control Assoc* 17:277-283.
10. Clarkson TW. 1995. *Am J Clin Nutr* 61:682S-6S.
11. Wolfe MF, Seiber JN. 1993. *Occup Med: State of the Art Rev.*
12. Winer AM, Atkinson R. In Long Range Transport of Pesticides; Kurtz DA, Ed.; Atmospheric reaction pathways and lifetimes for organophosphorous compounds,. Lewis Publishers. 1990. pp 115-126.
13. Woodrow JE, Crosby DG, Seiber JN. 1983. *Residue Rev* 85:111-125.
14. Hooper MJ, Detrich PJ, Weisskopf CP, Wilson BW. 1989. *Bull Environ Contam Toxicol* 42:651-659.
15. Eldred RA, Cahill TA, Wilkinson LK, Feeney PJ, Chow JC, Malm WC. 1990. Measurement of fine particles and their chemical components in the IMPROVE/NPS networks. From Visibility and Fine Particles, a Transaction of the Air and Waste Management Association, 187-196.
16. Zabik J.M. Atmospheric Transport and Deposition of Pesticides from California's Central Valley into the Sierra Nevada Mountains. University of California at Davis, Davis, CA, 1993.
17. Zabik JM, Seiber JN. 1993. *J Environ Qual* 22:80-90.
18. Cahill TA, Carroll JJ, Campbell D, Gill TE. Air Quality. In Status of the Sierra Nevada, Sierra Nevada Ecosystem Project, Final Report to Congress. 1993. Volume 2, 1227- 126.
19. Meylan W, Howard P. 1996. Melting point, boiling point, vapor pressure estimation program. Syracuse Research Corporation, Syracuse, NY, USA.
20. Aston LS. The Interaction Between Organophosphorous Pesticide Vapors and the Needles of *Pinus ponderosa*. University of California at Davis, Davis, CA, 1996.
21. Aston LS, Seiber JN. 1997. *J Environ Qual* 26: 1483-1492.
22. Schomburg CJ, Glotfelty DE, Seiber JN. 1991. *Environ Sci Technol* 25:155-160.
23. Seiber JN, Wilson BW, McChesney MM. 1993. *Env Sci and Tech* 27:2236-2243.
24. McConnell LL, LeNoir JS, Datta S, Seiber JN. 1998. *Environ Tox and Chem* 17:1908-1916.
25. Datta S, Hansen L, McConnell LL, Baker J, LeNoir J, Seiber JN. 1998. *Bull Environ Contam Toxicol* 60:829-836.
26. Stebbins RC. 1985. A Field Guide to Western Reptiles and Amphibians. Houghton Mifflin, Boston, MA.
27. Jennings ME. 1996. Status of Amphibians. Chapter 31 in Sierra Nevada Ecosystem Project: Final Report to Congress, Volume 2, pp. 921-944.

28. Bradford DF, Tabatabai F, Graber DM. 1993. *Conserv Biol* 7:882-888.
29. Drost CA, Fellers GM. 1996. *Conserv Biol* 10:414-425.
30. Bradford DF. 1991. *J Herpetol* 25: 174-177.
31. Bradford DF, Graber DM, Tabatabai, F. 1994. *Southwest Nat* 39:323-327.
32. Fellers GM. National Biological Services Division, USGS, personal communication, 1998.
33. Fulton MH, Chambers JE. 1985. *Toxicol Lett* 26:175-180.
34. Berrill M, Bertram S, McGillivray L, Kolohon M, Pauli B. 1994. *Environ Tox Chem* 13:657-664.
35. Schmidt WF, Hapeman CJ, Fetttinger JC, Rice CP, Bilboulian S. 1997. *J Agric Food Chem* 45:1023-1026.
36. Montgomery JH. 1993. Agrochemicals Desk Reference, Environmental Data. Lewis Publishers, Chelsea, Michigan.
37. Reijnders PJH. 1986. *Nature* 324:456-457.
38. Fry DM, Toone CK. 1981. *Sci* 213: 922-924.
39. Guillette LJ, et al. 1994. *Env Health Persp* 102:680-688.
40. Keith LH. Environmental Endocrine Disruptors. A Handbook of Property Data. John Wiley & Sons, Inc., New York, NY, 1997.
41. Ludke JL. 1977. *Pestic Biochem Physiol* 7:28-33.
42. Schuytema GS, Nebeker AV. 1996. Amphibian Toxicity Data for Water Quality Criteria Chemicals. EPA/600/R-96/124.
43. Berrill M, Coulson D, McGillivray L, Pauli B. 1998. *Environ Toxicol and Chem* 17:1738-1744.
44. Suntio LR, Shiu WY, Mackay D, Seiber JN, Glotfelty D. 1988. *Rev of Environ Contam and Toxicol* 103:1-59.
45. Kim YH. 1985. Evaluation of a Gas Chromatographic Method for Estimating Vapor Pressure with Organic Pollutants. Ph.D. dissertation, University of California, Davis.
46. Woodrow JE, Seiber JN, Baker LW. 1997. *Environ Sci Technol* 31:523-529.
47. Taylor AW, Spencer WF. 1990. Volatilization and Vapor Transport Processes. In, Pesticides in the Soil Environment. Chap. 7, SSSA, #2.
48. Wauchope RD, TM Butler, AG Hornsby, PWM Augustijn Beckers, JP Burt. 1992. The SCS/ARS/CES pesticide properties databases for environmental decision-making. *Rev Environ Contam Toxicol* 123:1-157.
49. Meausred by GC retention.
50. Woodrow JE, Honagonahalli P, Seiber JN. 1998. Calculated downwind pesticide air concentrations from estimated emissions and air dispersion models. Abstract 6E-004, 9th International Congress, Pesticide Chemistry (IUPAC), 2-7 August, London, UK.
51. Harris ML, Bishop CA, Struger J, Ripley B, Bogart JP. 1998. *Environ Toxicol and Chem* 17:1351-1363.
52. Herner, A.E. 1995. ARS PPD. Agricultural Research Service, Pesticide Properties Database. U.S. Department of Agriculture, Beltsville, MD.
53. State of California Pesticide Use Reports 1990-1996. Department of Pesticide Regulation, Sacramento, CA, USA.

Chapter 6

Instrumental and Bioanalytical Measures of Endocrine Disruptors in Water

Shane A. Snyder, Erin Snyder, Daniel Villeneuve, Kannan Kurunthachalam, Alex Villalobos, Alan Blankenship, and John Giesy

Department of Zoology, National Food Safety and Toxicology Center, and Institute for Environmental Toxicology, Michigan State University, East Lansing, MI 48824

The threat of modulation of animal endocrine systems by xenobiotic compounds has become a major issue facing researchers today. These compounds are varied in form and mechanism of action. This poses unique challenges in the identification and evaluation of these compounds from environmental matrixes. The aquatic environment is exceptionally susceptible to xenobiotic insult. Water can be polluted by a multitude of sources and acts as a sink for many types of pollution. This chapter outlines several methods for the detection and quantitation of endocrine disrupting compounds (EDCs) in the aquatic environment. However, no single method alone can predict or detect all EDCs present in an environmental sample, nor can all the biological mechanisms of action be accounted for in one simple test. Therefore, comprehensive screening for EDCs must combine several types of analyses including *in vivo* and *in vitro* bioassays, and analytical chemistry. Each type of analyses has advantages and disadvantages which will be discussed in this chapter.

A number of compounds released into the environment by human activities can modulate endogenous hormone activities and have been termed endocrine-disrupting compounds (EDCs) (1, 2). Environmental EDCs have been defined as exogenous agents which interfere with the "synthesis, secretion, transport, binding, action, or elimination of natural hormones in the body that are responsible for the maintenance of homeostasis, reproduction, development, and/or behavior" (3). It has been hypothesized that such compounds may elicit a variety of adverse effects in both humans and wildlife, including promotion of hormone-dependent cancers, reproductive tract disorders, and reduction in reproductive fitness (4-10). Endocrine "disruption" by environmental contaminants has become a cause for concern (11) and as a result there is an urgent need to develop methods for monitoring EDCs in the environment. This need is underscored by recent legislation mandating that chemicals and formulations be screened for potential estrogenic activity before they are manufactured or used in certain processes (Safe Drinking Water Act Amendments of 1995 - Bill Number S.1316; Food Quality Protection Act of 1996 - Bill Number P.L. 104-170).

EDCs act through a number of mechanisms including receptor-mediated hormonal responses such as hormone mimics (1). These include xenobiotic effects through thyroid hormone receptor, epidermal growth factor (EGF) receptor, aryl hydrocarbon receptor (AhR), estrogen receptor (ER) and androgen receptor (AR) (12, 13). Of all the EDCs, those that are direct-acting estrogen receptor (ER) agonists (xenoestrogens or estrogen mimics), direct acting ER antagonists (antiestrogens), or androgen antagonists have received the greatest attention, due in part to their importance in embryonic development. The "estrogen hypothesis" stemmed from the observation that some of the effects observed in oviparous wildlife exposed to persistent and bioaccumulative chemicals were similar to those that could be caused by injecting estrogen into eggs (2). This hypothesis was supported by the fact that naturally occurring exoestrogens, such as phytoestrogens, could cause reproductive dysfunction in animals. Further support came from the observation that some xenobiotics that can bind the ER are weak estrogen agonists or antagonists in *in vitro* expression assays (12). Because the major initial concern over endocrine modulating compounds has been the estrogen mimics (xenoestrogens), we will restrict our discussion to "estrogenic" and "antiestrogenic" compounds, but the techniques presented can be applied to any receptor-mediated process for which a suitable *in vitro* cell system can be developed (12).

There are a wide variety of ER-active compounds in the environment (14). These include both natural products (11) and synthetic compounds (10). The synthetic compounds include both chlorinated and non-chlorinated compounds that can either mimic or antagonize the effects of endogenous estrogen (12). Some of the compounds found in the surface waters that have been reported to be estrogenic include nonylphenol (NP), octylphenol (OP), 17β-estradiol (E2) (produced in the body) and the synthetic estrogen ethinylestradiol (EE2) (used in oral birth control medication) (15).

Several types of *in vitro* assays are available for measuring the estrogenic activity of single compounds or complex mixtures. The range of responses includes everything from simple receptor binding, to expression of endogenous or exogenous genes, to cell proliferation and differentiation. *In vitro* systems are attractive as screening tools because they are rapid, inexpensive, and fairly reproducible. For these reasons, precise estimates of the relative potency of a great many samples or compounds can be obtained in a rather short period of time. Expression assays examine induction or suppression of proteins encoded by genes whose transcription is thought to be modulated through an ER-mediated mechanism. Increases or decreases in the activity of the protein of interest upon exposure to a single compound or complex mixture, such as an environmental extract, suggest the presence of one or more ligands with the potential to modulate a broad range of genomically controlled estrogenic responses.

Although *in vitro* assays are an attractive option for screening and mechanistic studies, they may miss effects that would take place only in whole organisms. Some toxicants require metabolic activation through one or more pathways that occur *in vivo*, but not *in vitro*. Homeostatic controls and bioaccumulation generally are not simulated by *in vitro* testing systems. Toxicant effects can occur by different mechanisms in multiple tissues simultaneously. In addition, the same chemical exposure can result in very different responses in an animal depending upon its life stage, sex, and reproductive state. Fish are useful *in vivo* models for testing for effects of endocrine disrupting chemicals in water. Some species are extremely sensitive to such effects, and fish can act as integrators of responses to mixtures of toxicants that occur in the environment. In addition, effects related to growth and reproduction in fish are more easily related to population-level and ecological effects than are effects in *in vitro* systems. Fish can also be used in conjunction with *in vitro* testing systems to investigate mechanisms of toxicant action and to calibrate *in vitro* to *in vivo* responses.

The aquatic environment is especially susceptible to the effects of contaminants. Effluents from municipal and industrial wastewater treatment facilities, storm and agricultural run-off, and boating add many exogenous compounds to the aquatic ecosystem. Of the most frequently studied EDCs, butyltin compounds are next in abundance to alkylphenols in wastewater effluents. These compounds are also present in surface waters from many locations. The impact of tributyltin on the marine environment has received considerable attention during recent years because of its high toxicity to aquatic organisms (16-20). The use of TBT as an antifouling agent in paints applied on ships results in direct contact with aquatic environments. However, studies on the contamination of surface water (lakes and rivers) by butyltin compounds originating from the disposal of municipal wastewater are limited.

In the aquatic environment, instrumental analytical techniques are used to identify and quantify EDCs in sediments, tissues, and water. In some cases the toxicant is already known. In other cases, screening can be used to identify new toxicants based on bioassay-directed fractionation coupled with subsequent instrumental analyses. This is part of a toxicity identification and evaluation (TIE). By using bioassay-directed fractionation, compounds with certain mechanisms of action can be isolated and identified. A mass-balance can then be performed to see if all observed bioactivity can be accounted for by the compounds identified (21). This comprehensive approach allows for screening of various types of EDCs and may identify yet unknown compounds with significant bioactivity. The issue of environmental endocrine disruption is extremely complex. No single method will identify or explain all types of EDCs in the aquatic environment. Integrated approaches that combine *in vitro* and *in vivo* bioassays with analytical chemistry techniques are needed to screen for and assess the effects of EDCs in the environment.

In Vitro Bioassays

In vitro bioassays have an important role to play in an integrated approach to the study of endocrine disrupting compounds in the environment (12). *In vivo* studies in both the field and laboratory are critical for linking exposure to biologically relevant effects. They are, however, impractical for routine, high throughput screening, and characterization of individual compounds or environmental samples. Procedurally, *in vitro* bioassays are typically performed more quickly and at significantly less cost than *in vivo* studies. Use of simplified biological systems circumvents much of the inter-individual, seasonal, and temporal variability which can confound interpretation of *in vivo* responses. Additionally, *in vitro* bioassays avoid many of the complex socio-political and ethical issues associated with whole animal studies. Thus, *in vitro* bioassays are more amenable systems for routine use than *in vivo* assays. *In vitro* bioassays use simplified biological or biochemical systems to measure responses elicited by exposure to individual compounds or environmental samples. The responses measured are generally highly sensitive, integrated, and mechanistically-based. These properties allow *in vitro* bioassays to overcome many of the limitations of *in vivo* studies and instrumental analyses.

While instrumental analyses are essential for the identification and quantitation of compounds, they provide no information regarding the efficacy and potency of those compounds to modulate endocrine function. *In vitro* bioassays measure mechanistically-based biological responses. *In vitro* bioassays provide an integrated measure of the combined potency of all compounds in a sample. This is a practical advantage, since instrumental analysis of complex mixtures can be very expensive, difficult, and time consuming. Additionally, *in vitro* bioassays can detect compounds for which there are no analytical methods available and can be useful tools to measure potential additive and non-additive interactions between compounds. When the biological response being measured is highly sensitive, *in vitro* bioassays can also account for compounds which can exert a biological effect at concentrations less than analytical detection limits. 17β-estradiol and ethinylestradiol, for instance, can elicit significant responses in MVLN cells at concentrations less than 50 pM (≈ 15 ppt) (12), while sensitive instrumental analyses, such as HPLC fluorescence which has a detection limit of approximately 100 ng/mL (15) (≈ 350 nM) for these compounds. As a result, *in vitro* bioassays provide information which can complement instrumental analyses.

In vitro bioassays have been used widely for the study of potentially estrogenic compounds (12). Competitive receptor binding assays (12, 22-25), cell proliferation assays (26-30) and *in vitro* gene expression assays (12, 24, 29, 31-35) are among the most common *in vitro* approaches employed in the study of estrogenic and other endocrine disrupting compounds. The mechanistic foundation, advantages, and disadvantages of these approaches have been described elsewhere (12, 22, 36).

Work in our laboratory has primarily involved the use of the MVLN (MCF-7-luc) *in vitro* gene expression assay. MVLN cells are human breast carcinoma cells stably transfected with a luciferase reporter gene under control of estrogen responsive elements (EREs) of the *Xenopus* vitellogenin A2 gene (37, 38). When cells are exposed to a single compound or environmental sample, ER ligands can enter cells and bind the ER (Fig. 1). In MVLN cells, binding also upregulates expression of an exogenous luciferase reporter gene (Fig. 1). Upon addition of the appropriate substrate, luciferase catalyzes a light producing reaction. As a result, luciferase activity can be measured conveniently and with great sensitivity using a 96 well plate-reading luminometer. The measured luciferase activity is proportional to the sample's ability to modulate ERE-mediated gene expression.

Although gene expression in MVLN and other estrogen responsive cells is generally regarded to be ligand induced, there are also ligand-independent pathways which can modulate the transcriptional activity of the ER through activation of protein kinases (12). For example, treatment of cells with growth factors or agents that activate protein kinases in general, can cause the ER to be activated (without ligand binding) to bind to EREs and activate transcription. Complex interactions within mixtures are also possible. Investigations have included mixtures of estrogenic and antiestrogenic compounds (39) and mixtures of weak estrogens (12).

Use of the MVLN assay by our laboratory provides examples of useful applications of *in vitro* bioassays as part of an integrated approach to the study of endocrine disruptors. Screening with the MVLN bioassay has identified both individual compounds and complex mixtures extracted from environmental samples that were able to induce ERE-mediated gene expression (12, 15, 40-42). Active samples have been compared in terms of either their relative efficacy (the magnitude of response elicited; (15, 40, 41)) or relative potency (the mass, volume, or concentration of sample or compound needed to elicit some defined level of response; 1). Environmental samples from a diverse array of sources including surface waters (15, 43), wastewaters (15), sediments (40), and commercial fish diets (42) have been evaluated successfully. MVLN-based screening and relative potency or efficacy comparison has been used to focus analytical efforts on samples that show the greatest potential to exert estrogenic effects, identify field sites that warrant more extensive study, and profile the distribution of potentially estrogenic agents in relation to point sources, land use, and wastewater treatment processes. In cases where chemical fractionation has been coupled with MVLN screening, the results have facilitated the formulation and testing of hypotheses regarding which specific compounds or groups of compounds in an environmental sample which may be responsible for the *in vitro* responses observed (40). MVLN responses to individual compounds have been used to characterize the relative "estrogenic" potencies of compounds (12). Comparison of responses elicited by a compound in several *in vitro* bioassays based on different mechanistic assumptions can provide information regarding that compound's mechanism of action. Thus, *in vitro* bioassays can provide a great deal of useful information to aid in the assessment of endocrine disrupting compounds.

As with other approaches, *in vitro* bioassay responses must be interpreted with a clear understanding of their limitations. They can detect and provide an integrated measure of the ability of a compound or sample to elicit a mechanism-specific biological response. The relevance of that response for modeling or predicting effects *in vivo* is dependent on the relevance of the mechanistic

Figure 1. Mechanism for activation of the estrogen receptor (ER) in MVLN cells

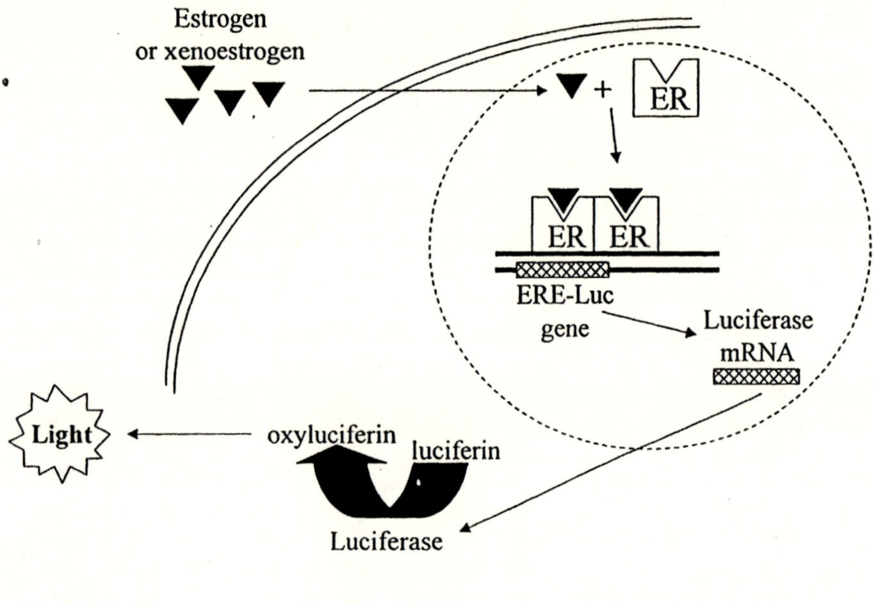

principles and assumptions upon which the assay is based. The MVLN assay, for example, would not detect compounds that exert estrogenic responses *in vivo* by acting as an anti-androgen or by altering endogenous estradiol levels through effects on the hypothalamus or pituitary. Thus, unless a general and fairly universal mechanism of action for a specific type of *in vivo* effect can be defined, *in vitro* bioassays provide a rather narrow field of view for detecting compounds which may alter endocrine function on a whole-organism level. Furthermore, because they use simplified systems, *in vitro* assays fail to account for a number of important factors which may modulate responses via a given mechanism of action *in vivo*. Such factors include toxicokinetics, cross-talk between biological pathways, and environmental factors (12, 22). In order to be useful for predicting effects at the whole-animal or population level, *in vitro* bioassays must be rigorously calibrated to *in vivo* responses (44, 45). The greatest limitation of *in vitro* bioassays for either screening or monitoring is the need for *a priori* knowledge about the mechanisms of action of a compound or mixture for which the assay is designed to screen. Thus, although *in vitro* bioassays have a role to play in the evaluation of endocrine disrupting compounds, they are just one of the tools needed to address the issue.

In vivo versus *in vitro* testing for EDCs

Despite public pressure to use *in vitro* testing whenever possible to spare animal lives and minimize the cost and time involved, whole animal studies are crucial for the development of biologically relevant methods to screen for and characterize EDCs in water. *In vitro* bioassays are mechanism specific and cannot account for the broad spectrum of potential mechanisms of effects that may be relevant *in vivo* (22). Whole animals can integrate and account for effects of chemicals in a mixture that may act through different mechanisms, at different tissues, either directly or indirectly to modulate overall endocrine function as well as specific endpoints. Additionally, effects of some compounds may be affected by metabolic transformation. Such transformations are generally not accounted for by *in vitro* bioassays (22). Furthermore, whole animal studies account for bioaccumulation and homeostatic controls that affect the potency or efficacy of compounds *in vivo*. Thus, although not necessarily amenable for routine monitoring and screening, *in vivo* studies are critical for relating exposure to population level-effects and ecologically relevant responses as well as the development and validation of relevant *in vitro* screening assays and biomarkers.

Use of fish to test for EDCs.

Fish have long been used as biosentinel models of water quality, particularly because of the concerns associated with the increased production, use, and disposal of chemicals in the environment (13, 46). Among these chemicals are the EDCs, which are thought to interfere (directly or indirectly) with the processes governing reproduction, behavior, and other aspects used to evaluate population health.

Many factors must be taken into account when using fish to test for effects of EDCs. It is well known that environmental factors play an important role in fish reproduction. Temperature, photoperiod, diet, availability of spawning substrates, proximity and reproductive readiness of potential mates (47), and water quality all affect reproductive status, ovulation, and spawning in fish (48). Stress, including that from capture, handling, and confinement, can have a rapid and marked effect on plasma sex hormone levels (49, 50). Seasonal and dailly changes in sex hormone levels must be considered in sampling schemes (49). If fish are fed during a study, diet must be selected carefully since some animal diets can contain hormonally active components or contaminants (30, 51, 52).

Fish exhibit a variety of reproductive strategies. Some are ovoviviparous and some are oviparous. Some, like the fathead minnow (*Pimephales promelas*), demonstrate parental care of eggs or offspring, while others, like the goldfish (*Carassius auratus*), ignore or even consume their eggs. Some spawn only once per breeding season, while others spawn repeatedly. This last point is particularly important to consider, since hormone levels and other end points can change rapidly and dramatically around the time of spawning.

Selection of test species. Different species of fish have been used to test for effects of EDCs, and each has its advantages and disadvantages. Desirable species should be easy to culture and handle in the laboratory, readily available year-round, widespread, and economically or aesthetically important. It is also important to choose a species for which there is a great deal of knowledge about its reproductive biology. It is difficult to determine whether an EDC has caused a departure from the normal condition if the normal condition is not known. Easily recognized sexual dimorphism is important for sexing fish, and in some cases, reduction of prominent secondary sex characteristics is used as evidence of endocrine disruption (44, 45). Some species that are currently used include trout and salmon, particularly rainbow trout (*Oncorhynchus mykiss*), carp (*Cyprinus carpio*), goldfish, white sucker (*Catostomus commersoni*), largemouth bass (*Micropterus salmoides*), fathead minnow, Japanese medaka (*Oryzias latipes*), mummichog (*Fundulus heteroclitus*), and mosquitofish (*Gambusia affinis*).

End points and biomarkers used to assess effects of EDCs on fish. Fish have been used in both laboratory and field studies to test individual chemicals or mixtures, effluents, and natural waters impacted by human activity for potential to alter reproductive structure and function. Some end points used to assess the effects of EDCs on fish and some studies that employed these end points are listed (Table 1). This table does not list all end points that could be used or all of the studies in which they have been used. For a review of methods to assess effects of EDCs on fish, see (53).

Plasma vitellogenin (Vtg) induction is a popular marker of exposure of fish to chemicals that bind to the estrogen receptor. Vtg is an egg-yolk precursor synthesized in the liver of oviparous animals in response to estrogen (54). Vtg is released from the liver to the bloodstream where it is carried to the ovary and sequestered into developing oocytes. Vtg usually is not detected in significant

TABLE 1: End points used to assess reproductive impacts of EDCs in Fish

End point	**Selected references**
vitellogenin (alkaline labile phosphorus, immunoassay, liver mRNA)	(56, 62)
gonadosomatic index (GSI)	(100-103)
plasma sex hormone levels (T, 11-KT, E2 and other estrogens)	(104, 105)
steroidogenesis enzyme activities and intermediates	(106)
production of hormones during *in vitro* organ incubation (gonad, pituitary)	(103, 105)
gonadotropins and gonadotropin releasing hormone	(103, 104)
gonad structure and histology (intersex, gonad duct development)	(107-109)
activity of liver enzymes involved in steroid metabolism (mixed function oxygenase, or MFO, activity)	(57, 110)
fertility and fecundity	(67, 102)
secondary sex characteristics	(44, 45)
reproductive behavior	(44, 111)

amounts in the plasma of male or immature fish, but they can be stimulated by exposure to natural or synthetic estrogenic chemicals to produce Vtg (55). Thus, Vtg is a useful tool to detect estrogenic activity of test chemicals or environmental waters potentially impacted by estrogenic EDCs. However, it is important to keep in mind that Vtg induction has not been shown to have any adverse effects in fish and is a functional measure of exposure useful only for monitoring for estrogenic compounds (55).

Field studies. Other researchers in the United States, the United Kingdom, and Canada have conducted field studies to determine whether EDCs in the environment have the ability to alter the reproductive structure and function of feral or wild (44, 56-60) and caged fish (61-63). Representative examples of these studies are those conducted with adult fish at the Aquatic Toxicology Laboratory at Michigan State University.

***In situ* exposure of caged fathead minnows to municipal waste water treatment plant effluents**. Fathead minnows were caged in rivers directly in the outfalls of municipal wastewater treatment plants (WWTPs) in mid-Michigan (64). Plasma Vtg, E2, testosterone (T), gonad histology, and male secondary sex characteristics were examined. Adult male and female minnows were caged for 3 weeks. No changes in any of these end points could be attributed to EDCs in WWTP effluents.

***In situ* exposure of caged common goldfish to municipal waste water treatment plant effluents.** This study was similar to the fathead minnow study previously described. Adult male and female goldfish were caged for 6 weeks at the same sites with the fathead minnows (65). Plasma Vtg, E2, and T were measured by enzyme immunoassay. Gonad histology was examined, and gonad tissue was removed and incubated *in vitro* to evaluate gonadal steroidogenic activity. Interpretation of the data is in progress.

Laboratory Studies. The Aquatic Toxicology Laboratory at Michigan State University has conducted both field and controlled laboratory studies. All of the laboratory studies were conducted with adult fathead minnows exposed to test compounds in a proportional flow-through diluter system.

Exposure of fathead minnows to 17ß-estradiol: plasma sex steroids, plasma Vtg, and egg production. Fathead minnows were exposed for 19 days to water borne E2 at measured concentrations ranging from 3.5 to 15 ng/L in a proportional flow-through diluter (66). Fish were exposed in duplicate groups of six, with three males and three females per exposure tank. Interestingly, E2 was detected even in control and solvent control tanks, indicating that the fish were a source for some of the estrogen in the exposure tanks. Egg production, plasma Vtg, plasma E2, and gonad histology were examined. The EC50 (concentration expected to cause 50% effect) for inhibition of egg production was 120 ng/L, and the EC50 for Vtg induction in males was 251 ng/L. Vtg was measured indirectly as alkaline labile phosphorus. In male fish, plasma E2 increased in a dose-dependent manner with concentrations of water borne E2. Plasma E2 levels in females were not affected by water borne E2 exposure. This difference in response may be due to sex-related differences in homeostatic control or metabolism of estrogen. Histopathological changes of the gonad were noted in a companion study (45) (see next section).

Exposure of fathead minnows to 17ß-estradiol: alterations in gonad histology and secondary sex characteristics. The fathead minnow was used as a test subject in part because of its prominent sexual dimorphism. Males have a fatpad (for which the fish is named) on the dorsal surface of the head. Males also develop nuptial tubercles, rows of small, hardened projections on the rostrum. The fatpad and nuptial tubercles are easily seen with the naked eye, and are absent in females. The female is generally lighter in color and smaller than the male, which usually develops a darker color pattern when sexually mature. In addition, the female has a more tapered head shape.

Sexually mature fathead minnows were exposed to water borne E2 as described in the previous section (66). In males, E2 exposure caused atrophy of secondary sex characteristics and histologic lesions of the testes. Lesions of the testes included proliferation of Sertoli cells, degenerative spermatozoa, and enlarged phagolysosomes containing necrotic spermatozoa and other cellular debris. Ovaries of females exposed to E2 contained more primary follicles and fewer secondary and Graafian follicles than ovaries of control fish (45), indicating that ovarian development in preparation for spawning had been retarded.

Exposure of fathead minnows to 4-nonylphenol. Adult fathead minnows were exposed to water borne 4-nonylphenol (NP) in a proportional flow-through diluter at concentrations ranging from <0.01 to 3.4 μg/L for 42 days (67). The study was conducted first early in the natural breeding season (Experiment I, or EX I) and again later in the breeding season (Experiment II, or EX II). Measured exposure concentrations for the two experiments were similar (EX I: 0.05, 0.16, 0.4, 1.1, and 3.4 μg NP/L ; EX II: <0.01, 0.10, 0.33, 0.93, and 2.4 μg NP/L). Examination of dose-response relationships between NP and egg production per female and between NP and plasma E2 in females revealed curves indicative of "Inverted-U" type responses, with lesser doses resulting in an increased response and greater doses resulting in lesser responses. For males and females in EX I, plasma E2 levels were greater than controls for all of the NP treatment groups except the greatest (3.4 μg NP/L), which was not different from the controls. Plasma E2 levels for both sexes ranged roughly from 2 to 25 ng/mL. For females in EX II, two intermediate exposure concentrations (0.1 and 0.33 μg NP/L) produced plasma E2 concentrations statistically greater than those in both controls groups. The plasma E2 concentrations for females ranged from about 2 to 8 ng/mL. In EX II males, none of the treatment levels produced plasma E2 levels statistically greater than both controls, but the sample sizes were small, and the shape of the curve indicated a trend similar to that seen in the females. For females in EX I, the dose-response curve for NP versus plasma Vtg suggests that NP caused a decrease in plasma Vtg levels, with the two greatest NP concentrations resulting in the least plasma Vtg levels. For females in EX II, the dose-response curve for NP versus plasma Vtg suggests an Inverted-U response, with the least NP exposure level producing a statistically significant increase over the controls, and greater exposure levels producing statistically lesser Vtg levels that are still significantly greater than the control Vtg levels. For males in both EX I and EX II, there was no significant increase in plasma Vtg in any of the NP exposure concentrations. In EX I, NP caused no statistically significant effect on egg production per female. However, the sample sizes used were small, and the shape of the dose-response curve suggests the same inverted-U type response, with increased egg production at the least treatment level (0.05 μg NP/L) and a decrease in greater treatment levels. In EX II, there was an increase in egg production per female for the greatest treatment level (2.4 μg NP/L) and for one of the intermediate treatment levels (0.01 μg NP/L), with two treatment levels between exhibiting no difference from the controls.

The mechanism of NP action in the alteration of fathead minnow reproductive end points is unclear, but may be related to positive and negative feedback on endogenous estrogen production. The differences in response over the two different experiments may have resulted from differences in the reproductive state of the fish. This underscores the importance of timing of the study relative to the reproductive state in fish. Furthermore, while NP has been reported to be a weak estrogen agonist, it is likely that the mechanism of action is not due to the binding of NP to the ER.

Exposure of fathead minnows to nonylphenol polyethoxylate (NPEO). Adult fathead minnows were exposed to an industrial mixture of NPEO for 42 days in a proportional flow-through diluter. Three males and three females were placed in each exposure tank. NPEO exposure concentrations ranged from 0 to 10 μg/L. Plasma Vtg, E2, and T were measured by enzyme immunoassay. Egg production per female (fecundity) was determined. No statistically significant concentration-dependent relationship was demonstrated between fecundity and NPEO exposure. However, the shape of the curve indicated a possible inverted-U response. NPEO was not observed to have any statistically significant effect on any of the other end points examined.

Use of the medaka fertilized egg nanoinjection assay as an alternative method of exposure in fish. The mechanisms playing a role in the disruption of endocrine, and related systems in fish need a better understanding (68). Considering that testing of EDCs in feral populations, or even representative captive adults in laboratory studies represent an economical, practical, and even technological obstacle, attention has been shifted to possible alternatives, not just of surrogate species, but also of exposure methodologies. The medaka (*Oryzias latipes*), a small fish frequently used as adults and/or early life stages (embryos and/or larvae) in aquatic toxicity studies (69-71) has proven useful as a tool in those alternative studies. The attributes that make this species attractive for *in vivo* experimentation have been discussed elsewhere (46); however, only until recently has micromanipulation technology started to take advantage of a more novel exposure approach: *in ovo* nanoinjection of chemicals into fertilized medaka eggs (72, 73).

While egg microinjecting techniques have been developed in several fish (mainly salmonids) over ten years ago (74-76), few studies have been attempted in species with small eggs. Recently, Villalobos *et al.* (73) reported success in using this medaka assay to test the toxicity of several technical mixtures of polychlorinated napthalenes (Halowax). The authors injected 0.5 nl of Halowax 1014, Halowax 1013, or Halowax 1051 in triolein as a carrier solvent. Mortality for the injected triolein controls was below 10%, and the Halowax-induced responses suggested involvement of the Ah receptor. The morphological effects of each of the Halowaxes was easily observed in the embryos and could be associated with structure-activity dependent effects (responses varied according to the chloronaphthalene speciation, chlorine localization within the molecule, and total mass contribution). Furthermore, the effects were consistently reproduced after repeated tests. However, these compounds did not appear to have elicited a disrupting effect in the medaka endocrine system. These studies, done in collaboration with the U.S. Geological Survey (USGS) Columbia Environmental Research Center (Columbia, MO), demonstrate the feasibility of nanoinjection of compounds into the small eggs of medaka.

Monitoring of butyltin compounds in water

Recently, widespread industrial and agricultural applications of organotin compounds, particularly butyltins, have led to increasing concerns about environmental contamination and biological effects caused by these compounds. Di- (DBT) and mono-butyltins (MBT) have been widely used as heat and light stabilizers for polyvinyl chloride (PVC) plastics and catalysts in the production of polyurethane and silicone. TBT has been used as antifouling agents, biocides, fungicides, molluscicides, insecticides and wood preservatives (77). The use of butyltin compounds in a variety of consumer products and household items (including sanitary napkins, diaper, cellophane wrap, sponges used in washing dishes, butter parchments, and gloves) (78, 79) and the leachates from PVC pipes have led to their occurrence in municipal wastewater and sewage sludge (80, 81). In addition, TBT is used as a disinfectant in waxes, polishes, sprays and laundry washes, which may allow contamination of municipal wastewater (82). Industrial discharges of TBT used as a slimicide in the paper industry and for textile and lumber treatment, and in cooling water treatment are further sources of contamination.

Input of DBT and TBT into surface waters from applications other than antifouling paint on vessels (i.e., use of TBT as a slimicide in the cooling water of a thermoelectric power plant in Italy) was first reported in the late 1980s (83). Concentrations of butyltins in raw wastewater in Zurich, Switzerland, on six sampling days ranged from 136 to 564, 127 to 1026 and 64 to 217 ng/L, respectively (84). Approximately 90% of each butyltin species was associated with suspended particles, and butyltins are removed at a rate of 62, 80 and 66% for MBT, DBT and TBT, respectively, by sedimentation (clarifier) in the sewage treatment plant. After secondary clarification, which usually represents the outflow from treatment plants, MBT, DBT and TBT ranged from 7 to 47 ng/L. MBT, DBT and TBT concentrations in the range of 0.1-0.97, 0.41-1.24 and 0.28-1.51 μg/g (dry wt) were found in sewage sludge. Similarly, MBT was found in all the 36 influent and effluent samples from WWTPs in five Canadian cities. Concentrations of MBT were in the range of 2.8-31 μg/L for influent and 1-22 μg/L for effluent. DBT and TBT were found only infrequently at a few μg/L (85). Concentrations of MBT, DBT and TBT in sewage sludge were as great as 650, 600, and 675 μg/g, dry wt, respectively (85). In Bordeaux, France, concentrations of MBT and DBT in influents from a WWTP were 18 and 12 ng/L, respectively (86), whereas those in effluent samples were 15 and 8 ng/L, respectively. TBT was not detected in these samples. Effluents from a WWTP in Mainz, Germany, contained MBT and DBT concentrations in the range of 13-28 and 6-96 ng/L, respectively (87). Total butyltin concentration in sewage sludge from Patna, India, was 340 ng/g, dry wt (88).

Studies describing the occurrence of butyltins in wastewater effluents in the U.S. are not available. Nevertheless, results from the above studies suggest that butyltins are widespread contaminants in wastewater. Even in those WWTPs with exceptionally modern systems (84), the concentrations of TBT remaining in effluents were greater than the environmental quality standard of 2 ng/L (89). Occurrence of butyltin compounds in unnavigable areas in certain rivers (90), as well as in aquatic organisms of certain unnavigable rivers, suggest the release of these compounds from wastewater (91).

Effects. Butyltin compounds, particularly TBT has been implicated in the development of "imposex" (development of male sex characteristics in females) in about 50 species of gastropods worldwide (92). This effect was attributed to inhibition by TBT of cytochrome P450-dependent aromatase, the enzyme responsible for the conversion of testosterone to 17β-estradiol (17). Since cytochrome P450 systems control the conversion of cholesterol into a variety of hormones including testosterone and estradiol, effects of TBT on this enzyme system can result in hormonal imbalance in animals. Although the concentrations of TBT in wastewater effluents were greater than those that could cause imposex in several sensitive gastropods, discharge of effluents into surface waters could result in dilution. Nevertheless, aquatic animals concentrate TBT in tissues, particularly in liver or hepatopancreas (20, 93, 94). Since lipovitellin (provides protein and lipid for the developing embryo) is produced in the hepatopancreas or liver, accumulation of butyltins in these organs can interfere with vitellogenesis (95). TBT can also interact with cytochrome P450 and modulate the toxic effects of contaminants such as polychlorinated biphenyls (96). Although MBT and DBT were not shown to induce imposex in gastropods, they have been reported to be toxic to rainbow trout yolk sac fry (97). No observed effect concentrations (NOEC) for DBT and TBT in water were 40 ng/mL and 40 pg/mL, respectively, for the mortality of rainbow trout yolk sac fry (97). Similarly, DBT has been reported to produce a wide variety of toxic effects such as immunosuppression in exposed fish (97, 98).

Instrumental analysis and biomonitoring of butyltins. The need to identify and determine concentrations of different butyltin species at trace levels was driven by their great toxicity. There are now a variety of instrumental methods available for the routine determination of butyltins in different matrices, but it should be noted that these analytical techniques are still somewhat laborious. The modern techniques are generally sensitive enough to measure butyltin species at sub ng/L levels, but quality assurance and quality control efforts should be implemented on a regular basis to assess the accuracy and precision of analytical results. Some of the recent instrumental analytical techniques for butyltins are discussed elsewhere (18). The most sensitive methods for analysis of butyltins involve the conversion to either alkyl derivatives or volatile hydrides followed by chromatographic separation and determination with specific detectors. The gas chromatography (GC) techniques used in butyltin analyses include: GC-MS (mass spectrometery), GC-GFAA (graphite furnace atomic spectrometry), and GC-FPD (flame photometric detector). Methods based on GC are widely used due to its high resolution and detector versatility.

Due to the potential for TBT to induce imposex in gastropods at water concentrations of a few ng/L, studies have suggested the application of field bioassays using transplanted bivalves to assess TBT contamination in aquatic systems (17, 92). The intensity of expression of imposex in *Nucella lapillus* can generally be related to the water concentration of TBT. The appearance of a small penis and the partial development of a vas deferens first occurs at TBT concentrations below 0.5 ng/L (as tin), although reproduction is unaffected at this level (89). At 1-2 ng/L (as tin) penis size is markedly increased and in some females proliferation of vas deferens tissue overgrows the genital papilla, thus sterilizing the animal. At slightly greater concentrations all females become sterile and a concentration of 10 ng tin/L caused suppression of oogenesis and initiated spermatogenesis (89). For monitoring purposes, the intensity of imposex is most simply described by the relative penis size (RPS) index in which the bulk of the female penis [$(\text{female length})^3/(\text{male length})^3 \times 100$] is expressed as a percentage of that of the male. Measurement of the length of a penis in species such as <u>Nucella</u> is a simple procedure since the penis can be expressed without dissection by placing the animal (minus shell) dorsal side up and drawing back the flap of tissue forming the rook of the mantle cavity. Under a binocular microscope and using 1 mm graduated graph paper, the length of the penis can be measured to the nearest 0.1 mm from its tip to its base. Bifurcate or trifurcate penises are occasionally found, particularly on females, and in these cases the longer or the longest structure is measured. A female lacking any measurable penile outgrowth is registered as zero and this value is included in the calculation of the mean. Whilst length is the most convenient parameter of penis size to measure, it does not convey a true impression of the difference in mass between, for example, a penis of 1 mm and another of 3 mm. However, the weight of the penis is related to the cube of its length (89) and thus the RPS calculation involves the cube of the length. An RPS of 50% indicates that the mean penis size of the female is half the bulk of that of the male. RPA is a reliable measure in all areas except those where contamination is severe. In these areas the male penis is often deformed and consequently the RPS value is lowered and imposex thereby underestimated. Since, RPS does not positively identify the presence of sterilized females and thus a second index, the vas deferens sequence (VDS) was devised (89). This index is based on the division of imposex into six stages ranging from early development to the final stage of sterilization. Calculation of the mean VDS stage provides an index by which to compare imposex in different populations. Any population with an index above 4 contains sterilized females and thus has a reduced reproductive capacity. Further details on this biomonitoring technique are given in Bryan et al., 1988. A positive relationship between the incidence of imposex and body residues of TBT has been shown (99). Although the induction of imposex tends to be regarded as a specific response to TBT, field bioassays can be confounded by several environmental factors and therefore their applicability to quantitatively monitor butyltin contamination is in question. *In vitro* bioassays have been used to study the effects of butyltin compounds on EROD activity. Due to high cytotoxic potential of butyltin compounds, use of *in vitro* studies to elucidate the mechanism of toxic effects of butyltin compounds has been hindered. This indicates the need for a combination of several techniques to identify the sources and effects of butyltins in the environment.

Bioassay-directed Chemical Fractionation

Unique challenges face the analytical chemist when trying to develop methods for EDCs. The primary difficulty is determining what compounds are EDCs. While various groups and agencies have published lists of EDCs, the question of new or unknown EDCs in the environment remains. By the use of *in vitro* bioassays like those previously discussed as an analytical tool, the chemist may begin to develop a more comprehensive screening program. When a bioactive sample is discovered, subsequent chemical fractionation and *in vitro* bioassays may lead to compounds or classes of compounds likely responsible for the observed effect. Once candidate compounds have been identified, a predicted activity can be estimated by applying relative potency factors derived from pure compounds. This predicted activity may be compared to the actual activity measured in the bioassay to determine if a mass-balance exists.

A reliable method for sensitive detection of certain EDCs in complex aqueous mixtures has been developed using solid-phase extraction (SPE) and *in vitro* bioassays (15). Dissolved and particulate phase organics were simultaneously extracted from 5 L water samples at the field site by pumping the samples sequentially through a 90 mm glass fiber filter and a 90 mm Empore styrenedivinylbenzene (SDB) SPE disk. The SDB matrix permits the extraction of both non-polar and semi-polar organics from water. Flow rates and volumes were measured with an in-line electronic flow meter/totalizer. Vacuum filtration was used at the laboratory to extract the SPE disks sequentially with acetone, dichloromethane and hexane. All extracts were dried over sodium sulfate and concentrated to near dryness. HPLC with a preparative silica column was used to generate three fractions of the extract based on polarity. The least polar fraction contained halogenated aromatic hydrocarbons (HAHs) and polycyclic aromatic hydrocarbons (PAHs), the moderately polar fraction contained alkylphenols and phthalates, and the fraction of greatest polarity contained steroids, bisphenol A, and alkylphenol polyethoxylate surfactants. Both total and fractionated extracts were assayed for estrogenic/antiestrogenic (ERE-mediated) activity or dioxin-like (AhR-mediated) activity using cell lines transfected with luciferase reporter genes under control of estrogen- or dioxin- responsive enhancer sequences, respectively. Additional fractionation by reverse phase HPLC and *in vitro* testing were sometimes used to more narrowly isolate bioactive compounds. Compounds identified using this strategy included: alkyphenols, steroids, PAHs, and HAHs. Instrumental analyses performed included: HPLC/fluorescence, GC/MS and GC/ECD. ELISA was used to quantify synthetic and endogenous estrogen in some bioactive extracts. Sites screened included several wastewater treatment plants in south central Michigan, effluents entering the Trenton Channel of the Detroit River, and areas of Lake Mead in Nevada.

The degree of induction over background was used as a measure of the activity of unknown samples to direct further fractionation and instrumental analyses. Most AhR-mediated activity occurred in the most non-polar fraction, while the greatest ERE-mediated activity occurred in the most polar of the three fractions. These polar fractions were further separated by reverse-phase HPLC into 9 fractions. In every case, the estrogenic activity occurred in the fractions containing E2 and EE2. These compounds were determined to be the most likely causes of observed *in vitro* estrogenicity. Further confirmation was completed using radioimmunoassays (RIAs) for E2 and EE2. Water concentrations of E2 and EE2 ranged from less than detection to 3.7 ng/L and 0.8 ng/L, respectively.

AhR-active compounds were detected by bioassay in only one sample, indicating that the levels of "dioxin-like" compounds were less than the method detection limits for most of the water samples. This is not suprising as only 5 L of water was extracted.

Although NP and OP, known environmental estrogens, were present in many water samples, there was no correlation between these alkylphenols and observed estrogenicity. Not only did the fraction containing these compounds show no bioactivity, the mass balance also would not predict a significant response based on the concentrations in the resulting extracts. Fine fractionation also confirmed that the lack of estrogenicity in the alkylphenol fraction was not due to antagonistic effects. The concentrations of NP and OP ranged from less than detection to 37 and 0.7 μg/L, respectively.

The *in vitro* bioassay systems were found to be effective tools for the screening of estrogen and dioxin-like activity in extracts of surface waters and for directing further fractionation and instrumental analyses. The results of the assay can also be compared to the predicted activity calculated from an additive model of estrogenic activity. In the second method, the concentrations of all of the putative estrogenic compounds in a mixture are multiplied by their relative potency factors (determined for individual compounds) and the total activity is predicted by summing the products of these two values. This method allows a mass balance of activity to be calculated. By comparing the response of the cells to whole extracts and various fractions, it is possible to determine whether all of the likely estrogenic compounds have been identified and to investigate nonadditive interactions among compounds.

Conlusions

Instrumental and bioanaltytical methods for the determination and measurement of EDCs in the aquatic environment have been presented. There are many tools available to aid in the identification and characterization of EDCs. *In vitro* bioassays have great potential for routine use in characterizing EDCs. They become particularly useful tools when used in conjunction with instrumental analyses. They are, however, only relevant as the mechanisms the are based on. Calibration to *in vivo* responses are crucial for proper application and interpretation of *in vitro* bioassay results. *In vivo* bioassays offer an intergrated measure of endocine disruption as the whole animal is taken into account. *In vivo* testing suffers from the complexity of sorting out mechanisms of toxicity. Analytical chemistry techniques are important in the identification and quantitation of EDCs in the environment and in controlled laboratory studies.

The issue of endocrine disruption is extremely complex. Only by combining various investigative techniques can the risks, if any, of EDCs be better understood. EDCs have a wide variety of toxicity mechanisms, while also having a great range of physical properties. An integrated approach is required to better detect, quantify, and understand EDCs in the environment.

References

(1) Kendall, R. J.; Dickerson, R. L.; Giesy, J. P.; Suk, W. A. *Principles and Processes for Evaluating Endocrine Disruptors in Wildlife*; SETAC Press.: Pensacola, FL, 1998.

(2) Feyk, L. L.; Giesy, J. P. In *Principles and Processes for Evaluating Endocrine Disruptors in Wildlife*; Kendall, R. J., Dickerson, R. L., Giesy, J. P., Suk, W. A., Eds.; SETAC Press.: Pensacola, FL, 1998, pp 121-140.

(3) U.S. EPA. **1997,** *Special Report on Environmental Endocrine Disruption: Effects Assesment and Analysis.* Office of Research and Development, Washington D.C. EPA/630/R-96/012.

(4) Safe, S. H. *Pharmacol. Ther.* **1995,** *67*, 247-281.

(5) Ahlborg, U. G.; al., e. *Critical Reviews in Toxicology* **1995,** *25*, 463-531.

(6) Sharpe, R. M.; Skakkebaek, N. E. *Lancet* **1993,** *341*, 1392-1395.

(7) Carlsen, E.; Giwercman, A.; Keiding, N.; Skakkebaek, N. E. *Environmental Health Perspectives* **1995,** *130*, 137-139.

(8) Horwitz, K. B.; Jackson, T. A.; Rain, D. L.; Richer, J. K.; Takimoto, G. S.; Tung, L. *Mol. Endocrinol.* **1996,** *10*, 1167-1177.

(9) Katzenellenbogen, J. A. *Environmental Health Perspectives* **1995,** *103*, 99-101.

(10) Gillesby, B. E.; Zacharewski, T. R. *Environmental Toxicology and Chemistry* **1998,** *17*, 3-14.

(11) Colborn, T. *Environmental Toxicology and Chemistry* **1998,** *17*, 1-2.

(12) Villeneuve, D. L.; Blankenship, A. L.; Giesy, J. P. In *Toxicant-Receptor Interactions and Modulation of Gene Expression*; Denison, M. S., Helferich, W. G., Eds.; Lippincott-Raven Publishers: Philadelphia, PA, 1997, pp 69-99.

(13) Giesy, J. P.; Snyder, E. M. In *Principles and Processes for Evaluating Endocrine Disruptors in Wildlife*; Kendall, R. J., Dickerson, R. L., Giesy, J. P., Suk, W. A., Eds.; SETAC Press: Pensacola, FL, 1998, pp 155-237.

(14) Kendall, R. J.; Brouwer, A.; Giesy, J. P. In *Principles and Processes for Evaluating Endocrine Disruptors in Wildlife*; Kendall, R. J., Dickerson, R. L., Giesy, J. P., Suk, W. A., Eds.; SETAC Press: Pensacola, FL, 1998, pp 1-16.

(15) Snyder, S. A.; Keith, T. L.; Verbrugge, D. A.; Snyder, E. M.; Gross, T. S.; Kannan, K.; Giesy, J. P. *Environ. Sci. Technol.* **1999** (in press).

(16) Alzieu, C. *Mar. Environ. Res.* **1991**, *32*, 7-17.

(17) Spooner, N.; Gibbs, P. E.; Bryan, G. W.; Goad, L. J. *Mar. Environ. Res.* **1991**, *32*, 37-39.

(18) Fent, K. *Crit. Rev. Toxicol.* **1996**, *26*, 1-117.

(19) Kannan, K.; Senthilkumar, K.; Loganahthan, B. G.; Takahashi, S.; Odell, D. K.; Tanabe, S. *Environ. Sci. Technol.* **1997**, *31*, 296-301.

(20) Kannan, K.; Guruge, K. S.; Thomas, N. J.; Tanabe, S.; Giesy, J. P. *Environ. Sci. Technol.* **1998**, *32*, 1169-1175.

(21) Sanderson, J. T.; Giesy, J. P. In *Encyclopedia of Environmental Analyses and Remediation*; Meyers, R. A., Ed.; John Wiley and Sons, Inc.: New York, NY, 1998, pp 5272-5297.

(22) Zacharewski, T. *Environmental Science & Technology* **1997,** *31*, 613-623.

(23) Kelce, W. R.; C.R., S.; S.C., L.; L.E., G.; J.A., K.; E.M., W. *Nature* **1995,** *375*, 581-585.

(24) White, R.; Jobling, S.; Hoare, A.; Sumpter, J. P.; Parker, M. G. *Endocrinology* **1994,** *135*, 175-182.

(25) Martin, P. M.; Horwitz, K. B.; Ryan, D. S.; McGuire, W. L. *Endocrinology* **1978,** *103*, 1860-1867.

(26) Soto, A. M.; C., S. *Biochemical and Biophysical Research Communications* **1984,** *122*, 1097-1103.

(27) Kelce, W. R.; E., M.; M.P., G.; S.C., L.; L.E., G. *Toxicology and Applied Pharmacology* **1994,** *126*, 276-285.

(28) Soto, A. M.; Chung, K. L.; Sonnenschein, C. *Environmental Health Perspectives* **1994,** *102*, 380-383.

(29) Olea, N.; Pulgar, R.; Pilar, P.; Olea-Serrano, F.; Rivas, A.; Novillo-Fertrell, A.; Pedraza, V.; Soto, A. M.; Sonnenshein, C. *Environmental Health Perspectives* **1996,** *104*, 298-305.

(30) Welshons, W. V.; Rottinghaus, G. E.; Nonneman, D. J.; Dolan-Timpe, M.; Ross, P. F. *Journal of Veterinary Diagnostic Investigations* **1990,** *2*, 268-273.

(31) Fluoriot, G.; Vaillant, C.; Salbert, G.; Pelissero, C.; Guirard, J. M.; Valotaire *Journal of Cell Science* **1993,** *105*, 407-416.

(32) Anderson, M. J.; Miller, M. R.; Hinton, D. E. *Aquatic Toxicology* **1996,** *34*, 327-350.

(33) Ramamoorthy, K.; Wang, F.; Chen, I. C.; Norris, J. D.; McDonnell, D. P.; Leonard, L. S.; Gaido, K. W.; Bocchinfuso, W. P.; Korach, K. S.; Safe, S. *Endocrinology* **1997,** *138*, 1520-1527.

(34) Miksicek, R. J. *Molecular Pharmacology* **1993,** *44*, 37-43.

(35) Kramer, V. J.; Helferich, W. G.; Bergman, A.; Klasson-Wehler, E.; Giesy, J. P. *Toxicology and Applied Pharmacology* **1997,** *144*, 363-376.

(36) Zacharewski, T. *Environmental Health Perspectives Supplements* **1998,** *106*, 577-582.

(37) Demirpence, E.; Duchesne, M. J.; Badia, E.; Gagne, D.; Pons, M. *Journal of Steroid Biochemistry and Molecular Biology* **1993,** *46*, 355-363.

(38) Pons, M.; Gagne, D.; Nicolas, J. C.; Mehtali, M. *Biotechniques* **1990,** *9*, 450-459.

(39) Blankenship, A.; Villeneuve, D.; Giesy, J. P. *Organohallogen Compounds* **1997,** *34*, 412-416.

(40) Khim, J. S.; Villeneuve, D. L.; Kannan, K.; Lee, K. T.; Snyder, S. A.; Koh, C. H.; Giesy, J. P. *Submitted to: Environ. Toxicol. Chem.* **1999**.

(41) Koistinen, J.; Soimasuo, M.; Tukia, K.; Oikari, A.; Blankenship, A.; Giesy, J. P. *Environmental Toxicology and Chemistry* **1998,** *17*, 1499-1507.

(42) Kemler, K.; Duda, C. A.; Blankenship, A. L.; Kannan, K.; Giesy, J. P. , Charlotte, NC 1998.

(43) Lee, K. T.; Villeneuve, D.; Kannan, K.; Giesy, J. P. , Charlotte, NC 1998.

(44) Bortone, S. A.; Davis, W. P. *BioScience* **1994,** *44*, 165-172.

(45) Miles-Richardson, S. R.; Kramer, V. J.; Fitzgerald, S. D.; Render, J. A.; Yamini, B.; Barbee, S. J.; Giesy, J. P. *Aquatic Toxicology* **1998,** *Submitted.*

(46) Villalobos, S. A.; Anderson, M. J.; Denison, M. S.; Hinton, D. E.; Tullis, K.; Kennedy, I. M.; Jones, A. D.; Chang, D. P. Y.; Yang, G.; Kelly, P. *Envir. Health Perspect.* **1996,** *104*, 735-743.

(47) Dulka, J. G.; Stacey, N. E.; Sorensen, P. W.; Van Der Kraak, G. J. *Nature (London)* **1987,** *325*, 251-253.

(48) Stacey, N. E.; Cook, A. F.; Peter, R. E. *Journal of Fish Biology* **1979,** *15*, 349-361.

(49) McMaster, M. E.; Munkittrick, K. R.; Van Der Kraak, G. J. *Canadian Technical Report of Fisheries and Aquatic Sciences* **1992,** *1836*, 1-29.

(50) Pankhurst, N. W.; Van Der Kraak, G.; Peter, R. E. *General and Comparative Endocrinology* **1995,** *99*, 249-257.

(51) Pelissero, C.; Le Menn, F.; Kaushick, S. *General and Comparative Endocrinology* **1991,** *83*, 447-457.

(52) Murphy, P. A.; Farmakalidis, E.; Johnson, L. D. *Food Chemistry and Toxicology* **1982,** *20*, 315-317.

(53) Donaldson, E. M.; Scherer, E. In *Methods for Assessing the Effects of Chemicals on Reproductive Functions*; Vouk, V. B., Sheehan, P. J., Eds.; John Wiley & Sons: New York, NY, 1983, pp 365-404.

(54) Mommsen, T. P.; Walsh, P. J. In *Fish Physiology*; CRC: Ann Arbor, 1988; Vol. XIA, pp 348-406.

(55) Sumpter, J. P.; Jobling, S. *Environmental Health Perspectives* **1995,** *103*, 173-178.

(56) Folmar, L. C.; Denslow, N. D.; Rao, V.; Chow, M.; Crain, D. A.; Enblom, J.; Marcino, J.; Guillette, L. J., Jr. *Environmental Health Perspectives* **1996,** *104*, 1096-1101.

(57) McMaster, M. E.; Van Der Kraak, G. J.; Portt, C. B.; Munkittrick, K. R.; Sibley, P. K.; Smith, I. R.; Dixon, D. G. *Aquatic Toxicology* **1991,** *21*, 199-218.

(58) Goodbred, S. L.; Gilliom, R. J.; Gross, T. S.; Denslow, N. P.; Bryant, W. B.; Schoeb, T. R. ; United States Department of the Interior. United States Geological Survey.: Sacramento, California, 1997, pp 48.

(59) Bevans, H. E.; Goodbred, S. L.; Miesner, J. F.; Watkins, S. A.; Gross, T. S.; Denslow, N. D.; Schoeb, T. ; United States Department of the Interior. United States Goological Survey.: Carson City, Nevada., 1996, pp 47.

(60) Lye, C. M.; Frid, C. L. J.; Gill, M. E.; McCormick, D. *Marine Pollution Bulletin* **1997,** *34*, 34-41.

(61) Harries, J. E.; Sheahan, D. A.; Jobling, S.; Matthiessen, P.; Neall, P.; Routledge, E. J.; Rycroft, R.; Sumpter, J. P.; Tylor, T. *Environmental Toxicology and Chemistry* **1996,** *15*, 1993-2002.

(62) Harries, J. E.; Sheahan, D. A.; Jobling, S.; Matthiessen, P.; Neall, P.; Sumpter, J. P.; Tylor, T.; Zaman, N. *Environmental Toxicology and Chemistry* **1997,** *16*, 534-542.

(63) Purdom, C. E.; Hardiman, P. A.; Bye, V. J.; Eno, N. C.; Tyler, C. R.; Sumpter, J. P. *Chemistry and Ecology* **1994,** *8*, 275-285.

(64) Nichols, K. M.; Miles-Richardson, S. R.; Snyder, E. M.; Giesy, J. P. *Environmental Toxicology and Chemistry* **1999,** *Submitted.*

(65) Snyder, E. M.; Nichols, K. M.; Pierens, S. L.; Snyder, S. A.; Giesy, J. P. , 1999.

(66) Kramer, V. J.; Miles-Richardson, S.; Pierens, S. L.; Giesy, J. P. *Aquatic Toxicology* **1998,** *40*, 335-360.

(67) Giesy, J. P.; Pierens, S. L.; Snyder, E. M.; Miles-Richardson, S.; Kramer, V. J.; Snyder, S. A.; Nichols, K. M.; Villeneuve, D. A. *Environmental Toxicology and Chemistry* **1998,** *Submitted.*

(68) Arcand-Hoy, L. D.; Benson, W. H. *Env. Toxicol. Chem.* **1998,** *17*, 49-57.

(69) Wisk, J. D.; Cooper, K. R. *Environ. Toxixol. Chem.* **1990**, 1159-1169.

(70) Marty, G. D.; Wetzlich, S.; Nuñez, J. M.; Craigmill, A.; Hinton, D. E. *Aquat. Toxicol.* **1991**, 329-340.

(71) Harris, G. E.; Metcalfe, T. L.; Metcalfe, C. D. *Environ. Toxicol. Chem.* **1994,** *13*, 1393-1403.

(72) Papoulias, D. M.; Nolte, D.; Tillitt, D. *Submitted to: Aquat. Toxicol.* **1999**.

(73) Villalobos, S. A.; Papoulias, D. M.; Meadows, J.; Blankenship, A. L.; Pastva, S. D.; Kannan, K.; Hinton, D. E.; Tillitt, D. E.; Giesy, J. P. *Submitted to: Environ. Toxicol. Chem.* **1999**.

(74) Black, J. J.; Maccubbin, A. E.; Schiffert, M. *J. Natl. Cancer Inst.* **1985**, 1123-1128.

(75) Metcalfe, C. D.; Cairns, V. W.; Fitzsimons, J. D. *Aquat. Toxicol.* **1988**, 347-356.

(76) Norrgren, L.; Andersson, T.; Bjork, M. *Aquat. Toxicol.* **1993**, 301-316.

(77) Piver, W. T. *Environ. Health Perspect.* **1973**, *4*, 61-79.

(78) Yamada, S.; Fujii, Y.; Mikami, E.; Kawamura, N.; Hayakawa, J.; Aoki, K.; Fukaya, M.; Terao, C. *J. AOAC Int.* **1993**, *76*, 436-441.

(79) Takahashi, S.; Mukai, H.; Tanabe, S.; Sakayama, K.; Miyazaki, T.; Masuno, H. *Submitted to: Environ. Pollut.* **1999**.

(80) Wu, W.; Roberts, R. S.; Chung, Y.-C.; Ernst, W. R.; Havlicek, S. C. *Arch. Environ. Contam. Toxicol.* **1989**, *18*, 839-843.

(81) Quevauviller, P. H.; Bruchet, A.; Donard, O. F. X. *Appl. Organomet. Chem.* **1991**, *5*, 125-129.

(82) Maguire, R. J. *Water Poll. Res. J. Canada* **1991**, *26*, 243-360.

(83) Bacci, E.; Gaggi, C. *Mar. Pollut. Bull.* **1989**, *20*, 290-292.

(84) Fent, K.; Muller, M. D. *Environ. Sci. Technol.* **1991**, *25*, 489-493.

(85) Chau, Y. K.; Zhang, S.; Maguire, R. J. *Sci. Total Environ.* **1992**, *121*, 271-281.

(86) Donard, O. F. X.; Quevauviller, P. H.; Bruchet, A. *Wat. Res.* **1993**, *27*, 1085-1089.

(87) Schebek, L.; Andreae, M. O.; Tobschall, H. J. *Environ. Sci. Technol.* **1991**, 871-878.

(88) Kannan, K.; Senthilkumar, K.; Sinha, R. K. *Appl. Organomet. Chem.* **1997**, *11*, 223-230.

(89) Bryan, G. W.; Gibbs, P. E.; Burt, G. R. *J. Mar. Biol. Ass. U.K.* **1988**, *68*, 733-744.

(90) Maguire, R. J.; Chau, Y. K.; Bengert, G. A.; Hale, E. J.; Wong, P. T. S.; Kramar, O. *Environ. Sci. Technol.* **1982**, *16*, 698-702.

(91) Kannan, K.; Grove, R. J.; Senthilkumar, K.; Henny, C. J.; Giesy, J. P. *Arch. Environ. Contam. Toxicol. (in press)* **1999**.

(92) Gibbs, P. E.; Bryan, G. W.; Pascoe, P. L. *Mar. Environ. Res.* **1991**, *32*, 79-87.

(93) Kannan, K.; Yasunaga, Y.; Iwata, H.; Ichihashi, H.; Tanabe, S.; Tatsukawa, R. *Arch. Environ. Contam. Toxicol.* **1995**, *28*, 40-47.

(94) Kannan, K.; Senthilkumar, K.; Elliott, J. E.; Feyk, L. A.; Giesy, J. P. *Arch. Environ. Contam. Toxicol.* **1998**, *35*, 64-69.

(95) Lee, R. F. *Mar. Environ. Res.* **1993**, *35*, 181-187.

(96) Kannan, K.; Villeneuve, D. L.; Blankenship, A. L.; Giesy, J. P. *J. Toxicol. Environ. Health* **1998**, *55*, 373-384.

(97) deVries, H.; Penninks, A. H.; Snoeij, N. J.; Seinen, W. *Sci. Total Environ.* **1991**, *103*, 229-243.

(98) Boyer, I. J. *Toxicol.* **1989**, *55*, 253-298.

(99) Horiguchi, T.; Shiraishi, H.; Shimizu, M.; Yamazaki, S.; Morita, M. *Mar. Pollut. Bull.* **1995**, *13*, 402-405.

(100) Pereira, J. J.; Mercaldo-Allen, R.; Kuropat, C.; Luedke, D.; Sennefelder, G. *Archives of Environmental Contamination and Toxicology* **1993,** *24*, 427-431.

(101) Jobling, S.; Sheahan, D.; Osborne, J. A.; Matthiessen, P.; Sumpter, J. P. *Environmental Toxicology and Chemistry* **1996,** *15*, 194-202.

(102) Monosson, E.; Fleming, W. J.; Sullivan, C. V. *Aquatic Toxicology* **1994,** *29*, 1-19.

(103) Singh, P. B.; Kime, D. E.; Epler, P.; Chyb, J. *Journal of Fish Biology* **1994,** *44*, 195-204.

(104) Van Der Kraak, G. J.; Munkittrick, K. R.; McMaster, M. E.; Portt, C. B.; Chang, J. P. *Toxicology and Applied Pharmacology* **1992,** *115*, 224-233.

(105) MacLatchy, D. L.; Van Der Kraak, G. J. *Toxicology and Applied Pharmacology* **1995,** *134*, 305-312.

(106) MacLatchy, D.; Peters, L.; Nickle, J.; Van Der Kraak, G. *Environmental Toxicology and Chemistry* **1997,** *16*, 1895-1904.

(107) Gimeno, S.; Komen, H.; Venderbosch, P. W. M.; Bowmer, T. *Environmental Science & Technology* **1997,** *31*, 2884-2890.

(108) Jobling, S.; Nolan, M.; Tyler, C. R.; Brighty, G.; Sumpter, J. P. *Environmental Science & Technology* **1998,** *32*, 2498-2506.

(109) Gray, M. A.; Metcalfe, C. D. *Environmental Toxicology and Chemistry* **1997,** *16*, 1082-1086.

(110) Johnson, L.; Casillas, E.; Sol, S.; Collier, T.; Stein, J.; Varanasi, U. *Marine Environmental Research* **1993,** *35*, 165-170.

(111) Kindler, P. M.; Bahr, J. M.; Philipp, D. P. *Hormones and Behavior* **1991,** *25*, 410-423.

Chapter 7

Potential Endocrine Disrupting Organic Chemicals in Treated Municipal Wastewater and River Water

Larry B. Barber[1], Greg K. Brown[1], and Steven D. Zaugg[2]

[1]U.S. Geological Survey, 3215 Marine Street, Boulder, CO 80303
[2]U.S. Geological Survey, P.O. Box 25046, MS 407, Denver, CO 80225

Select endocrine disrupting organic chemicals were measured in treated wastewater from Chicago, IL, Minneapolis/St. Paul, MN, Detroit, MI, and Milwaukee, WI, and in the Des Plaines, Illinois, and Minnesota Rivers during the fall of 1997 and the spring of 1998. Emphasis was given to alkylphenolpolyethoxylate (APEO) derived compounds, although 17-β-estradiol, bisphenol A, caffeine, total organic carbon, ethylenediaminetetraacetic acid (EDTA), and other compounds also were measured. Contaminants were isolated by continuous liquid-liquid extraction (CLLE) with methylene chloride and analyzed by gas chromatography/mass spectrometry in full scan and selected ion monitoring modes. The extracts were derivatized to form the methyl esters of alkylphenolethoxycarboxylates (APEC), and EDTA was isolated by evaporation and derivatized to form the tetrapropyl ester. The mass spectra of nonylphenol (NP) and octylphenol (OP) compounds are complex and show variations among the different ethoxylate and carboxylate homologs, reflecting variations in the ethylene oxide chain length. Recoveries for target compounds and surrogate standards ranged from 20-130%, with relative standard deviations of 9.9-53%. Detection limits for the various compounds ranged from 0.06-0.35 μg/L. Analysis of the wastewater effluents detected a number of compounds including NP, NPEO, OP, OPEO, NPEC, caffeine, and EDTA at concentrations ranging from <1-439 μg/L, with EDTA and NPEC being most abundant. There was variability in compound distributions and concentrations between the various sewage treatment plants, indicating differences in treatment type and influent composition. Several wastewater-derived compounds were detected in the river samples, with EDTA and NPEC persisting for considerable distance downstream from wastewater discharges, and NP and NPEO being attenuated more rapidly.

The impact of endocrine disrupting chemicals (EDC) in the environment is a topic of growing concern (*1-4*). An important source of EDC to hydrological systems is treated municipal and domestic wastewater, and there is growing evidence that disposal of wastewater effluent into surface waters may have an adverse effect on the endocrine system of aquatic organisms (*5-10*). Municipal wastewater is a complex mixture of natural and synthetic organic chemicals and potentially contains thousands of compounds. Some of the synthetic organic compounds that are common in treated wastewater have been shown in laboratory studies to induce endocrine disrupting effects. These compounds include natural hormones such as estrogen, degradation products of nonionic surfactants, and plasticizers. Nonionic surfactants are widely used synthetic organic chemicals that find their way into natural waters from domestic, industrial, and agricultural practices. An important group of nonionic surfactants are alkylphenolpolyethoxylates (APEO), which consist of an alkylphenol (AP) moiety, typically nonylphenol (NPEO) and octylphenol (OPEO), with 1-100 ethylene oxide (EO) units attached in the para position (most cleaning products have 9-10 EO units). The 1988 US production of APEO nonionic surfactants exceeded 450 million pounds, of which 55% was used in industrial applications, 30% in institutional cleaners, and 15% in household and personal care products (*11*). When APEO surfactants are discharged into the aquatic environment, biodegradation and sorption are the primary removal processes, although photolysis and volatilization also can be important (*12-14*). The APEO surfactants undergo biodegradation during wastewater treatment to produce short chain APEO (1-4 EO), alkylphenolethoxycarboxylates (APEC), and alkylphenols (AP).

The occurrence and fate of AP compounds in the environment have received considerable attention due to their persistence and aquatic toxicity (*11, 15, 16*), and recently have been reviewed (*17*). Several of the APEO compounds are listed on the U.S. Environmental Protection Agency, Toxic Substance Control Act Priority Testing list (*18*). Recent interest has focused on the effects of AP, APEO, and APEC on the endocrine system (*19-24*). The most common surfactant formulations use nonylphenol (NP) and octylphenol (OP). Technical grade OP occurs primarily as a single isomer (95% 4-1,1,3,3-tetramethylbutylphenol) because it is synthesized from phenol and isobutylene (*25*). In contrast, technical grade NP is an isomeric mixture of branched nonyl (C_9) structures produced through acid catalyzed alkylation of phenol with nonene (*25*). In addition to the alkyl-chain isomers, technical NP is a mixture of positional isomers with approximately 3-6% 2-NP and 90-96% 4-NP. The 2-NP isomers have shorter chromatographic retention times than the 4-NP isomers, but have similar isomeric C_9 side chains. Technical NP mixtures also contain 1-5% OP and decylphenol.

A number of other alkylphenol compounds are used as antioxidants. In particular the tert-(*t*)-butyl derivatives such as 2,6-di-*t*-butylphenol (DTBP), 2,6-di-*t*-butyl-4-methylphenol (BHT), and 2[3]-*t*-butyl-4-methoxyphenol (BHA) are widely used antioxidants in the polymer and food industries (*25*). The antioxidants can be oxidized (*26*) to their benzoquinone derivatives (i.e., DTBP and BHT oxidize to 2,6-di-*t*-butyl-1,4-benzoquinone, DTBB). A variety of alkylphenol antioxidants have been identified in wastewater and septage (*27, 28*), wastewater-contaminated groundwater (*28, 29*), and effluents from chemical manufacturing facilities (*30*).

Wastewater-derived alkylphenols have been extensively studied. Giger and others (*31*) measured NP and NP1EO-NP4EO by steam distillation and gas chromatography/mass spectrometry (GC/MS), and determined total concentrations in sewage effluents from Zurich ranging from 100-300 μg/L. Stephanou and Giger (*32*) measured NP and NP1EO-NP4EO using liquid/liquid extraction (LLE) with methylene chloride followed by GC/MS analysis. The combined concentrations of NP, NP1EO, and NP2EO in sewage effluents from Zurich ranged from <10-200 μg/L, NP3EO and NP4EO were not detected, and OP2EO was identified in several effluents. These concentrations were measured prior to a voluntary ban on NPEO surfactants in laundry detergent formulations, which has resulted in a 10-fold decrease in concentrations in present-day Swiss effluents (*33*). Naylor and others (*34*) measured NP and NPEO (NP1EO-NP17EO) in the Fox River system of Wisconsin, and found that NP concentrations in treated municipal wastewater ranged from 0.83-14 μg/L, and in treated pulp mill effluent ranged from <0.02-21 μg/L. Concentrations of total NPEO in treated municipal wastewater ranged from 47-350 μg/L, and in treated pulp mill effluent ranged from 3.1-730 μg/L. Concentrations of NP in river water ranged from 0.12-0.64 μg/L, whereas total NPEO concentrations ranged from 0.97-8.2 μg/L. Paxeus and others (*35*) analyzed municipal wastewater treatment plant influent and effluent in Sweden between 1989 and 1991 using LLE with methylene chloride and GC/MS. Although NP was the most abundant alkylphenol, the wastewater also contained other compounds such as BHT and BHA. The mean concentrations of NP in sewage influent decreased from 75 μg/L in 1989 to 9 μg/L in 1991 in response to source control measures implemented in 1989. Concentrations in treated effluent were <1 μg/L. Lee and Peart (*36*) measured NP and OP in treated wastewater and sludge by GC/MS as the acetyl derivatives. Concentrations of NP in wastewater from Toronto ranged from 0.8-30 μg/L and OP ranged from 0.1-2.5 μg/L. Bennie and others (*37*) used similar methods to measure NP, OP, and NP1EO-NP2EO in 35 surface waters of the Great Lakes basin and Upper St. Lawrence River. Measurable quantities of NP and OP were found in 24% of the samples (NP ranged from <0.01-0.92 μg/L and OP from <0.01-0.08 μg/L), NP1EO was detected in 58% of the samples (<0.02-7.8 μg/L), and NP2EO was detected in 32% of the samples (<0.02-10 μg/L). A number of high performance liquid chromatography (HPLC) techniques have been used for analysis of AP and APEO (*38-42*). The primary advantage of HPLC is the ability to measure the higher EO homologs. However, HPLC typically does not resolve individual NP and OP isomers or NPEO and OPEO homologs, and has the potential for interfering components. In this investigation we focused our effort on GC/MS techniques.

Ahel and others (*43*) determined nonlyphenolethoxycarboxylates (NPEC) by LLE with chloroform at pH 2 and gaseous stripping with nitrogen into ethyl acetate, fractionation on silica, derivatization with boron trifluoride (BF_3)/methanol or hydrochloric acid/methanol, and analysis by GC/MS and HPLC. They applied the method to treated wastewater in Switzerland and found total NP1EC-NP2EC concentrations ranging from 16-330 μg/L, considerably higher than NP (3-57 μg/L) and NP1EO-NP2EO (21-254 μg/L). Clark and others (*44*) measured APEO and APEC in large volume (500 L) samples of finished drinking water from New Jersey extracted by on-line continuous LLE with methylene chloride. Analysis by GC/MS identified NP1EO (0.08 μg/L) and NP2EO (0.15 μg/L), whereas HPLC/MS analysis identified OP2EO-OP8EO (0.13 μg/L), NP3EO-NP7EO (0.50 μg/L), OP2EC-OP4EC (0.05 μg/L), and NP2EC-NP7EC (0.23 μg/L). Field and Reed (*45*) developed a method for analyzing NPEC based on anion-exchange solid phase extraction, derivatization with methyl iodide to form the methyl esters, and analysis by GC with flame ionization detection and positive chemical ionization MS with ammonia as reagent gas. They found total NPEC concentrations in the Fox River, WI area ranging from 140-270 μg/L in treated municipal wastewater, from 17-1,200 μg/L in pulp mill effluent, and from <0.10-13.5 μg/L in river water. Fujita and others (*46*) applied the method of Schaffner and Giger (*47*), which forms the propyl esters of APEC, to effluent from an advanced wastewater treatment plant in Los Angeles and measured APEC concentrations of about 5 μg/L.

The impact of endogenous and synthetic sex hormones such as 17-β-estradiol and 17-α-ethynylestradiol in wastewater on fish reproductive development has been reported (*8, 10*). Desbrow and others (*8*) analyzed treated wastewaters in the United Kingdom for sex hormones and found 17-β-estradiol concentrations ranging from 0.003-0.048 μg/L, estrone concentrations ranging from 0.001-0.076 μg/L, and 17-α-ethynylestradiol concentrations ranging from <0.001-0.007 μg/L. Studies on natural urinary and synthetic ovulation inhibiting steroid hormones in raw and treated wastewater in the US (*48*) showed that secondary treatment removed 50-70% of the natural compounds and 20-40% of the synthetic compounds. Concentrations of total 17-β-estradiol (free and conjugated forms after hydrolysis) ranged from <0.01-0.02 μg/L, estrone ranged from <0.01-0.04 μg/L, and 17-α-ethynylestradiol ranged from 0.25-1.8 μg/L in treated wastewater.

Another phenolic EDC that is present in treated wastewater is bisphenol A, a chemical intermediate used to make epoxy resins, polycarbonate plastics, flame retardants, and dental sealents (*49*). Bisphenol A has been shown to have about 500 times the estrogenic activity of OP (*50*). Rudel and others (*28*) reported bisphenol A in septage, untreated wastewater, and treated wastewater at concentrations ranging from 0.04-1.7 μg/L.

In this study we apply an analytical method for APEO derived compounds to 8 sewage treatment plant effluents and 3 receiving streams during two seasons. The objective was to determine the quantity of total alkylphenolic EDC present in the various effluents and receiving streams. We also briefly discuss the receiving water assimilation capacity and seasonal variations in concentrations.

Methods

Field Sampling. Grab samples were collected during October 1997 and February/March 1998 from the sites listed in Table I. Treated municipal wastewater and river water samples were collected in 1-L amber glass bottles (one raw, one raw preserved with 1% formalin, v/v) and stored at 4° C until analysis. Eight sewage treatment plants were selected to evaluate different types of treatment ranging from secondary to advanced tertiary. Three receiving streams were evaluated: (1) the Minnesota River above the Minneapolis/St. Paul, MN wastewater discharges, (2) the Illinois River upstream of Peoria, IL which is about 20% wastewater, and (3)

Table I. Locations and characteristics of sites sampled during this investigation. Bold-italicized letters in parentheses are abbreviations used in headings of following tables. [MGD, million gallons per day; NA, not applicable; 7Q10, 7-day low flow for a 10 year period in cubic feet per second, CFS; L&D, lock and dam; RM, river mile]

Wastewater Plant/River Name	Location	Discharge Point	NPDES Permit No.	Treatment Type	Design Flow in MGD (Population served)	Upstream 7Q10 in CFS
Illinois River (***IR***)	Ottawa, IL	Starved Rock L&D RM 231	NA	NA	NA	2,310[a]
Des Plaines River (***DR***)	Joliet, IL	Brandon Road L&D RM 286	NA	NA	NA	1,281[a]
Minnesota River (***MR***)	Shakopee, MN	Minnesota River RM 25	NA	NA	NA	123[b]
Stickney (***ST***)	Cicero, IL	Chicago Sanitary & Ship Canal RM 315-317	IL0028053	Activated sludge secondary treatment with tertiary nitrogen removal	1,200[c] (4,000,000)	400[a]
Northside (***NS***)	Skokie, IL	North Shore Channel RM 331+	IL0028088	Activated sludge secondary treatment with tertiary nitrogen removal	333[c] (2,500,000)	0.14[a]
Calumet (***CA***)	Chicago, IL	Little Calumet River RM 321	IL0028061	Activated sludge secondary treatment with tertiary nitrogen removal	354[c] (2,000,000)	21[a]
Seneca (***SE***)	Eagan, MN	Minnesota River RM 6	MN0030007	Activated sludge secondary treatment with tertiary nitrogen removal	38[b] (245,000)	168[d]
Metropolitan (***ME***)	St. Paul, MN	Mississippi River RM 835	MN0029815	Activated sludge secondary treatment with tertiary nitrogen removal	251[b] (1,919,400)	1,019[d]
Blue Lake (***BL***)	Shakopee, MN	Minnesota River RM 20	MN0029882	Activated sludge secondary treatment with tertiary nitrogen removal	42[b] (252,000)	123[d]
Jones Island (***JI***)	Milwaukee, WI	Lake Michigan via Milwaukee Outer Harbor	W003682-1	Activated sludge secondary treatment	208[e] (555,000)	NA
Detroit (***DE***)	Detroit, MI	Detroit River	MI0022802	Secondary oxygenated activated sludge treatment	859[f] (3,150,000)	18,863[g]

[a] Illinois State Water Survey (1993), Contract Report 545.
[b] Kent Johnson, Minneapolis/St. Paul Metropolitan Wastewater Treatment Plant, personal communication.
[c] As specified in NPDES permit, values in million gallons per day (MGD).
[d] Gary Kimball, Minnesota Pollution Control Agency, personal communication.
[e] John Moser, Milwaukee Metropolitan Sanitation District, personal communication.
[f] Steve Kuplicki, Detroit Water and Sewage, personal communication.
[g] Flow exceeded 95% of the time.

the Des Plaines River upstream of the confluence with the Illinois River which is wastewater dominated.

Analytical Methods. The samples were analyzed for ammonium and total organic carbon (TOC) as gross indicators of treatment performance. Ammonium was measured spectrophotometrically and TOC was measured by persulfate/ultraviolet light oxidation and conductivity measurement.

The analytical method for the nonpolar, polar, and ionic wastewater derived compounds listed in Table II used continuous liquid-liquid extraction (CLLE) followed by derivatization and GC/MS analysis. The ionic strength of the raw 1-L sample was increased by adding 60 g/L sodium chloride, and the sample was extracted for 6 hours with 70 mL of methylene chloride. Extraction was accomplished by continuously refluxing solvent through a coarse glass frit at the top of the sample, causing formation of micro-droplets that travel down through the sample matrix. An elongated extraction chamber increases the surface area/volume ratio for the solvent, resulting in high extraction efficiency.

The APEC compounds require derivatization of the carboxylic acid functional group in order to be analyzed by GC/MS. After initial GC/MS analysis for the nonionic compounds, the CLLE residues were reacted with 1-mL of BF_3/methanol (10% v/v) for 5 minutes at 60° C to form the methyl esters of APEC (*43*). The reaction mixture was cooled, distilled water added, and the methylated acids extracted into hexane. The hexane was passed through a column of sodium sulfate and evaporated to a final volume of 500 µL followed by GC/MS analysis. The APEC compounds also were determined by a modification of the method of Schaffner and Giger (*47*) for nitrilotriacetic acid (NTA), which also determines ethylenediaminetetraacetic acid (EDTA). A 100-mL formalin-preserved sample was evaporated to dryness (90° C for ~24 hours), formic acid was added, and the sample was vacuum evaporated to dryness. The residue was reacted with 10% (v/v) acetyl chloride/1-propanol at 85° C for 1 hour to form the propyl esters of APEC, NTA, and EDTA. The esters were extracted into chloroform, the chloroform was evaporated to dryness, the residue was redissolved in 100 µL of toluene, and the extract was analyzed by GC/MS.

Analysis of steroid compounds (as well as APEO) by GC/MS is enhanced by derivatizing the hydroxyl functional groups (*51-54*). The CLLE extracts were reacted for 5 hours at 90° C with 25-µL of bis-(trimethylsilyl)trifluoroacetamide containing 10% trimethylchlorosilane. This reaction forms the trimethylsilyl (TMS) ethers of the hydroxyl functional groups and also derivatizes the carboxylate functional group of APEC compounds to form the TMS esters. Some of the steroid compounds such as estrone have a keto functional group that does not form a TMS derivative. The keto group can be reacted with methoxyamine hydrochloride to form the oxyamine, and then the hydroxyl groups derivatized to the TMS ethers.

The various extracts were analyzed by electron impact GC/MS using a Hewlett Packard 5890 GC with a 5970 Mass Selective Detector in both the full scan and selected ion monitoring (SIM) modes. The analyses were carried out under the following conditions: column, Hewlett Packard Ultra II, 5% phenylmethyl silicone, 25 m x 0.2 mm i.d., 33 µm film thickness; carrier gas, ultra high purity helium at 27 cm/second; splitless mode, purge valve on at 0.75 minutes; injector

Table II. Summary of CLLE spike and recovery data for compounds (listed in gas chromatographic elution order) measured in this study. Also shown are the molecular ions [M^+] and qualifier ions (Q_1, Q_2,) used in the SIM GC/MS analysis. [Rec, recovery; SD, standard deviation; RSD, relative standard deviation; value in parentheses is number of replicates; compound abbreviations used in text and tables are shown in bold-face italicized letters; IS, internal standard; SS, surrogate standard; nd, not determined]

Compound	[M^+] (*m/z*)	Q_1 (*m/z*)	Q_2 (*m/z*)	Rec[a] %	SD %	RSD %
4-methylphenol (7)	107.2	77.2	53.1	77	11	14
4-ethylphenol (3)	122.3	107.2	77.1	85	nd	nd
d_8-napthalene IS	136.2	108.3	68.2	nd	nd	nd
4-propylphenol (3)	136.3	107.2	77.2	79	nd	nd
4-*t*-butylphenol (3)	150.3	135.3	107.2	75	nd	nd
4-*t*-pentylphenol (3)	164.4	135.3	107.2	81	nd	nd
2,6-di-*t*-butylphenol (7), ***DTBP***	206.3	191.3	57.2	32	8.1	26
2,6-di-*t*-butyl-1,4-benzoquinone (7), ***DTBB***	220.3	177.3	135.2	120	12	9.9
2[3]-*t*-butyl-4-methoxyphenol (7), ***BHA***	180.2	165.2	137.1	23	4.9	21
d_{10}-acenaphthene IS	164.2	162.4	80.2	nd	nd	nd
d_{21}-2,6-di-*t*-butyl-4-methylphenol SS (7), ***d_{21}-BHT***	241.5	222.4	66.2	86	20	23
2,6-di-*t*-butyl-4-methylphenol (7), ***BHT***	220.3	205.3	57.2	20	4.0	20
1-phenylnonane IS	204.3	92.1	105.1	nd	nd	nd
4-*t*-octylphenol (7), ***OP***	206.3	135.1	107.2	76	12	16
4-nonylphenol[b] (7), ***NP***	220.3	135.2	107.1	73	11	15
4-*n*-octylphenol, ***n-OP***	206.3	135.1	107.1	nd	nd	nd
d_{10}-phenanthene IS	188.3	160.1	80.2	nd	nd	nd
caffeine (7)	194.2	109.2	82.2	74	13	17
4-*n*-nonylphenol SS (11), ***n-NP***	220.2	107.1	77.1	100	20	20
$^{13}C_6$ 4-*n*-nonylphenol SS (11), ***$^{13}C_6$ n-NP***	226.4	113.1	82.1	110	19	17
4-*t*-octylphenol monoethoxylate (3), ***OP1EO***	250.4	179.2	135.1	74	nd	nd
4-nonylphenol[b] monoethoxylate (7), ***NP1EO***	264.7	179.2	193.1	74	9.2	12
$^{13}C_6$ 4-*n*-nonylphenol monoethoxylate SS (11), ***$^{13}C_6$ n-NP1EO***	270.4	157.2	113.1	130	23	18
4-*t*-octylphenol diethoxylate (3), ***OP2EO***	294.4	223.2	135.2	80	nd	nd
d_6-bisphenol A SS (7)	234.2	216.2	121.1	110	15	14
bisphenol A (7)	228.2	213.2	119.2	96	22	23
4-nonylphenol[b] diethoxylate (7), ***NP2EO***	308.8	223.6	237.4	63	6.3	10
$^{13}C_6$ 4-*n*-nonylphenol diethoxylate SS (11), ***$^{13}C_6$ n-NP2EO***	314.5	113.1	201.1	66	32	48
4-*t*-octylphenol triethoxylate (3), ***OP3EO***	338.5	267.2	135.2	42	nd	nd
d_{12}-chrysene IS	240.3	236.3	120.2	nd	nd	nd
4-nonylphenol[b] triethoxylate (3), ***NP3EO***	352.4	267.7	281.3	64	nd	nd
estrone (3)	270.2	185.2	213.2	60	nd	nd
17-α-estradiol SS (11)	272.2	213.2	160.2	95	16	17
17-β-estradiol (7)	272.2	213.2	160.2	35	12	33
$^{13}C_6$ 4-*n*-nonylphenol triethoxylate SS (11), ***$^{13}C_6$ n-NP3EO***	358.5	133.1	226.2	nd	nd	nd
4-*t*-octylphenol tetraethoxylate (3), ***OP4EO***	382.5	311.4	133.0	nd	nd	nd
17-α-ethynylestradiol (3)	296.3	213.2	159.1	68	nd	nd
4-nonylphenol[b] tetraethoxylate (3), ***NP4EO***	397.0	311.7	325.3	60	nd	nd
d_{12}-perylene IS	264.3	260.3	132.5	nd	nd	nd
3-β-coprostanol (7)	388.5	233.3	215.2	38	14	38
cholesterol (7)	386.5	147.2	301.2	34	18	53
4-nonylphenol[b] monoethoxycarboxylate[c] (5), ***NP1EC***	292.4	207.1	193.0	88	30	34
4-nonylphenol[b] diethoxycarboxylate[c] (5), ***NP2EC***	336.5	117.1	251.1	60	28	47
4-nonylphenol[b] triethoxycarboxylate[c] (5), ***NP3EC***	380.5	295.3	309.1	100	36	36
4-nonylphenol[b] tetraethoxycarboxylate[c] (5), ***NP4EC***	424.6	117.1	339.3	100	53	53
4-bromophenylacetic acid[c] SS, ***BPAA***	228.1	169.0	171.0	nd	nd	nd
ethylenediaminetetraacetic acid[d], ***EDTA***	460.6	230.3	144.1	nd	nd	nd
d_{12}-ethylenediaminetetraacetic acid[d] SS, ***d_{12}-EDTA***	472.6	236.3	150.2	nd	nd	nd

[a] Distilled water spiked with 0.5 μg/L of target compounds and surrogate standards, extracted with 70-mL methylene chloride, and reduced to a final volume of 200 μL. Concentrations calculated from a 6-point calibration curve using deuterated PAH internal standards at 2 ng/μL.

[b] Isomeric mixture.

[c] Methyl ester derivatives.

[d] Propyl ester derivatives.

temperature, 300° C; initial oven temperature, 40° C; initial hold time, 1 minute; ramp rate, 6° C/minute; final temperature, 300° C; final hold time, 15 minutes; source temperature, 250° C; source pressure, 1 x 10^{-5} torr; ionization energy, 70 electron volts; MS interface temperature, 280° C; scan range 40 to 550 amu; scan time, ~1 scan/second.

Several surrogate standards were used for AP and APEO analysis. During the fall 1997 sampling, deuterated BHT (d_{21}-BHT) and bisphenol A (d_6-bisphenol A) obtained from Cambridge Isotope Laboratories (CIL, Andover, MA) were used. During the spring 1998 sampling, the surrogate standards for NP and NPEO were 4-normal-nonylphenol (4-*n*-NP) from Dr. Ehrenstorfer GmbH (Augsburg, Germany), and ring-labeled $^{13}C_6$ 4-*n*-NP, $^{13}C_6$ 4-*n*-NP1EO, $^{13}C_6$ 4-*n*-NP2EO, and $^{13}C_6$ 4-*n*-NP3EO (CIL). The surrogate standard for the NPEC methyl esters was 4-bromophenylacetic acid (BPAA, Aldrich, Milwaukee, WI), and for the NPEC and EDTA propyl esters was d_{12}-EDTA (CIL). The surrogate standard for the steroids was 17-α-estradiol (Aldrich). All surrogate standards were added to the sample prior to isolation to evaluate whole method recovery. Deuterated polyaromatic hydrocarbons (PAH) internal standards (Supelco, Bellefonte, PA) were added to the extracts prior to GC/MS analysis. Target compounds were spiked into distilled water and processed through the entire procedure to determine recoveries. Quantitation was based on deuterated PAH standards and response factors for the individual target compounds determined from a 6-point calibration curve. Recoveries for NP, NPEO, NPEC, bisphenol A, and steroid compounds were evaluated using the surrogate standards.

Results

Compound Characterization. Technical grade NP standards representing several CAS numbers were characterized, and the isomeric distributions of the various standards were nearly identical. Technical nonylphenol standards from Aldrich and Schenectady International (no CAS numbers given) have been extensively characterized (*55, 56*) and GC/MS analysis typically resolves 12-18 isomers (Figure 1). Although most of the NP and OP compounds are available as high purity standards, pure homolog NPEO, OPEO, and NPEC compounds are not readily available. The lower EO chain NPEO standards used in this study were technical mixtures containing multiple homologs (Figure 2). Analysis by flow injection mass spectrometry (Colleen Rostad, USGS, personnel communication) showed that the short chain NPEO standard (Witconol NP-15) was a relatively simple mixture (~90% NP1EO-NP4EO and 10% other impurities), whereas the longer chain standard (Witconol NP-40) was more complex (26% NP2EO-NP4EO, 63% NP5EO-NP11EO, and 11% other impurities). The GC/MS method could only determine NP1EO-NP5EO in Witconol NP-40, which contained up to NP11EO. The NPEC compounds are degradation products of NPEO surfactants, and well-characterized standards are not readily available. The NPEC standards used in this study were prepared by Union Carbide and are described in Field and Reed (*45*). Each of the NP1EC-NP4EC standards was a mixture of homologs and other impurities as determined by flow injection mass spectrometry (Figure 3).

Nonylphenol elutes as a cluster of isomers (Figure 1a) with diagnostic mass spectra (Figure 1b-f) containing a molecular ion [M^+] at *m/z* 220, and a homologous mass series corresponding to $HOC_6H_4(CH_2)_n$ (*m/z* 107, 121, 135, 149, 163, 177, and 191) with base peaks at *m/z* 107, 121, 135, and 149. The relative ion distributions for the different NP isomers vary substantially, reflecting branching on the C_9 side chain. Wheeler and others (*56*) classified the 4-NP spectra into 5 groups containing primary (Ar-CH_2-), secondary (Ar-CHR_1-), and tertiary (Ar-CR_1R_2-) structures. The primary α-carbon isomers form the *m/z* 107 base peak via benzylic cleavage with little additional fragmentation. However, the *m/z* 107 ion is common to all NP isomers and can be produced by simple cleavage or other mechanisms such as formation of a hydroxyl tropylium ion. Due to complex rearrangements, the *m/z* 107 ion is not useful in assigning structure. Secondary α-carbon isomers have a hydrogen atom and a methyl group attached to the benzyl carbon, and undergo fragmentation to form the α-methyl carbocation (*m/z* 121). Tertiary carbon compounds produce spectra with *m/z* 135 as the base peak representing the α,α-dimethyl carbocation which is very stable. Group 1 isomers (Figure 1b) consist of α,α-dimethyl structures with a primary or secondary β-carbon. The *m/z* 135 ion results from rupture at the benzylic bond with loss of C_6H_{13} [M^+ - 85]. The group 2 isomers (Figure 1c) have a base peak at *m/z* 121 [M^+ - C_7H_{15}] and a major ion at *m/z* 149 [M^+ - C_5H_{11}], with abundant ions at *m/z* 107, 135, and 191. These isomers have α-methyl-α-ethyl structures with a primary or secondary β-carbon. The group 3 isomers (Figure 1d) have a base peak at *m/z* 149 and no major ions between *m/z* 151-220. The *m/z* 149 base peak is similar to the α-methyl-α-ethyl structure of group 2, but the lack of a significant ion at *m/z* 191 indicates that the ethyl group is not attached to the α-carbon. A structure with a methyl group on the β-carbon and a primary γ-carbon will cleave to the β-carbon and have a high abundance of *m/z* 149. Subsequent loss of the β-carbon plus the methyl group forms the *m/z* 121 ion. The group 4 isomers (Figure 1e) have a base peak at *m/z* 107, and major ions at *m/z* 121 and 163 (cleavage at the α-carbon and loss of C_4H_9 from the β-carbon), indicating a α-methyl structure with a secondary β-carbon. The group 5 isomers (Figure 1f) have a base peak at *m/z* 121 and major ions at *m/z* 163 and 177 representing structures with a tertiary α-methyl-α-propyl-α-butyl configuration. One of the group 5 isomers is the first to elute in the NP mixture, indicating a highly branched structure that results in a lower boiling point. Octylphenol elutes as a single isomer (4-1,1,3,3-tetramethylbutylphenol) and has a mass spectrum similar to group 1 NP isomers with a base peak at *m/z* 135 reflecting the α,α-dimethyl configuration (data not shown) but with [M^+] at *m/z* 206.

The NP1EO-NP4EO homologs elute as isomer clusters similar to NP, and the number of resolvable isomers decreases with increasing EO units (Figure 2a). The NPEO compounds are more difficult to analyze than NP because the EO functional groups enhance interactions with active sites in the GC/MS system, resulting in selective loss and lower response factors for the NP3EO and higher homologs. Homologs greater than NP5EO are not readily analyzed by GC/MS due to decreasing volatility as well as greater activity. The mass spectra of NP1EO-NP4EO (Figure 2b-e) have [M^+] at *m/z* 264, 308, 352 and 396 representing addition of one (*m/z* 44), two (*m/z* 88), three (*m/z* 132), and four (*m/z* 176) EO units to NP. The mass spectra shown are for the first eluting NP isomer (group 5). This isomer

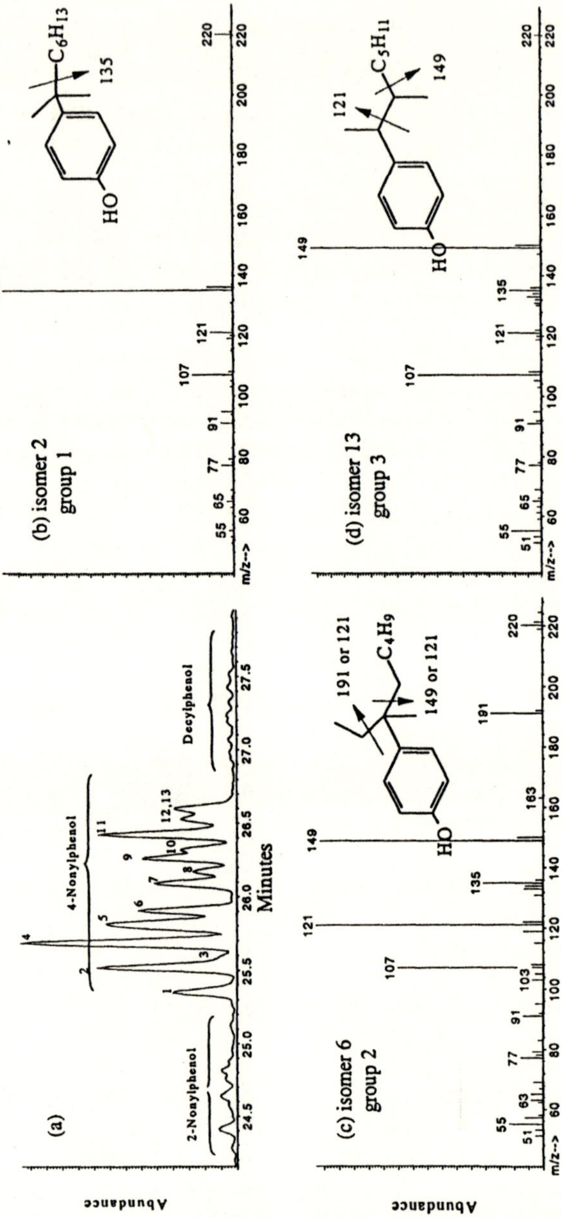
(a)
2-Nonylphenol
4-Nonylphenol
Decylphenol
Minutes
Abundance
(b) isomer 2
group 1
C_6H_{13}
135
(c) isomer 6
group 2
C_4H_9
191 or 121
149 or 121
(d) isomer 13
group 3
C_5H_{11}
121
149
HO
m/z-->
Abundance

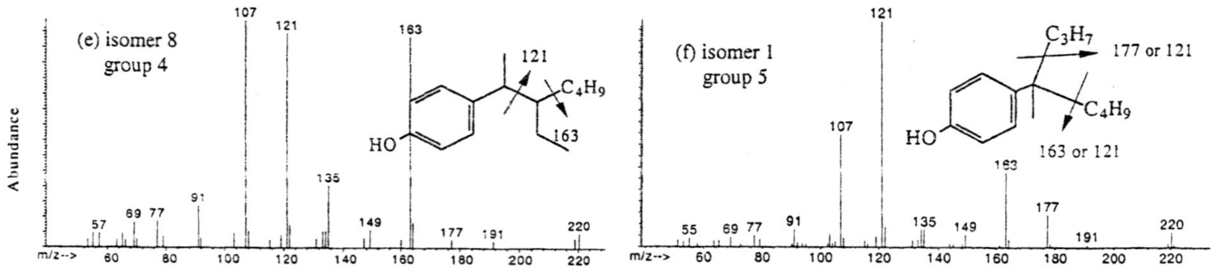

Figure 1. Full scan total ion chromatogram (a) and selected mass spectra (b-f) for Schenectady International high purity (95%) 4-nonylphenol [CAS# 84852-15-3*, asterisk denotes mixed isomers] standard (50 ng/μL). See text for group discussion.

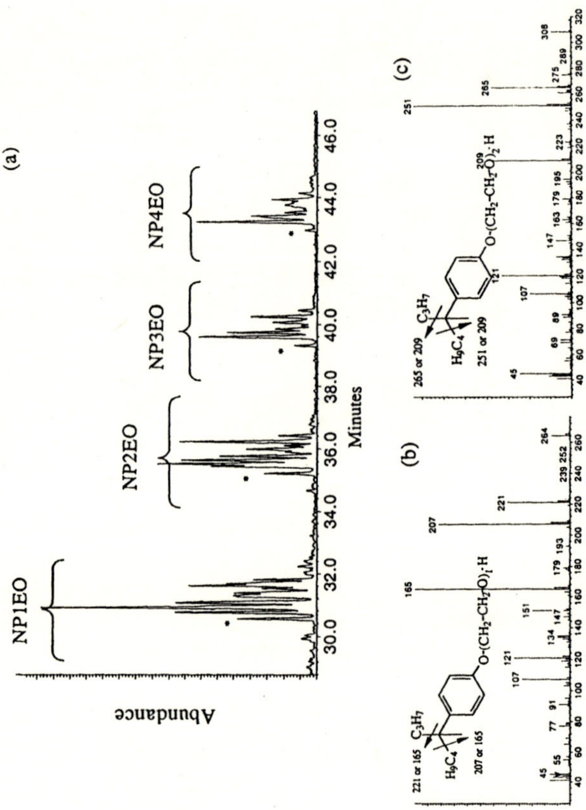
(a)
NP1EO
NP2EO
NP3EO
NP4EO
Abundance
Minutes
(b)
(c)

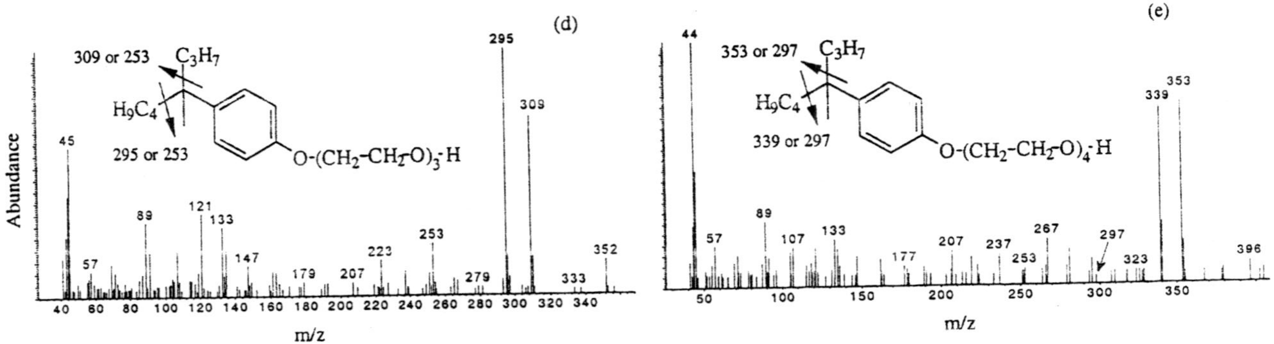

Figure 2. Full scan total ion chromatogram (a) and select mass spectra (b-e) for NP1EO-NP4EO. [Standard is a 50 ng/μL 50:50 mixture of Witconol NP-15 (5.1% NP1EO, 43.8% NP2EO, 31.6% NP3EO, 7.9% NP4EO) and Witconol NP-40 (0.0% NP1EO, 1.8% NP2EO, 7.8% NP3EO, 16.4% NP4EO); mass spectra are for the first group 5 peak (denoted by *) in the NP1EO-NP4EO isomeric cluster]

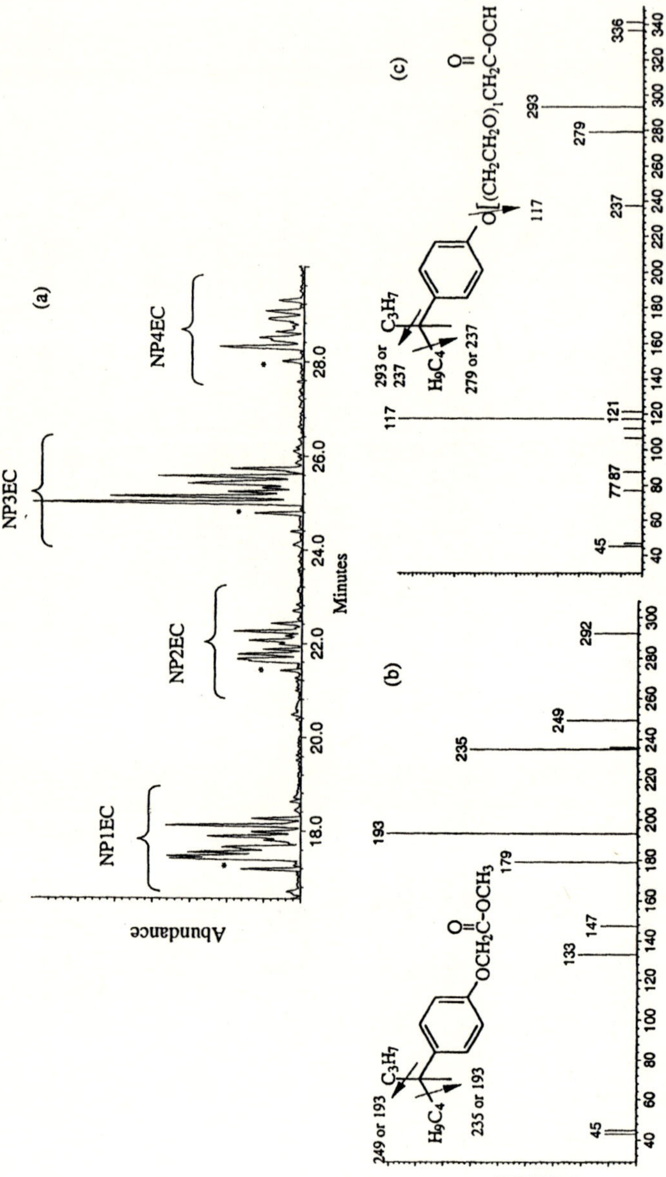
(a)
NP1EC
NP2EC
NP3EC
NP4EC
Abundance
Minutes
18.0
20.0
22.0
24.0
26.0
28.0
(b)
249 or 193
C_3H_7
H_9C_4
235 or 193
OCH$_2$C-OCH$_3$
Abundance
45
133
147
179
193
235
249
292
(c)
293 or 237
C_3H_7
H_9C_4
279 or 237
O[(CH$_2$CH$_2$O)$_1$CH$_2$C-OCH
117
45
77
87
121
237
279
293
336

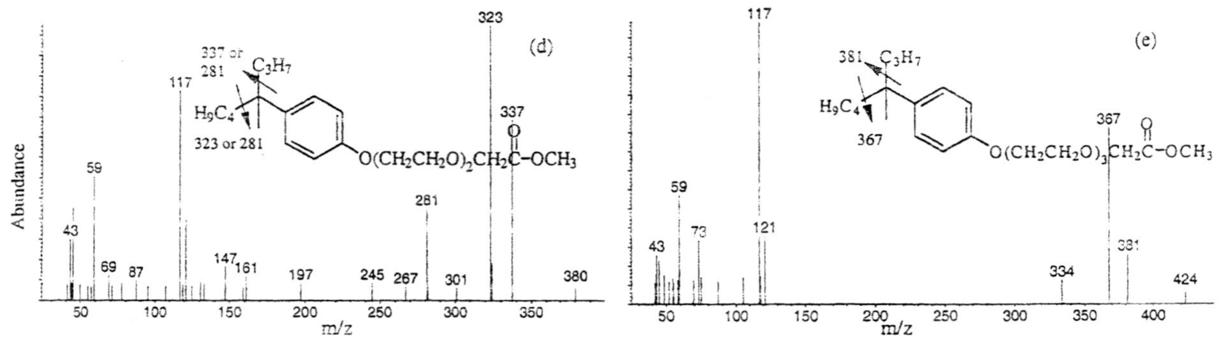

Figure 3. Full scan total ion chromatogram (a) and select mass spectra (b-e) for the methyl esters of NP1EC-NP4EC. [Standard is a 50 ng/µL mixture of NP1EC (90.0% NP1EC, 0.2% NP2EC, 0.5% NP3EC, 0.0% NP4EC), NP2EC (23.8% NP1EC, 45.0% NP2EC, 1.2% NP3EC, 0.3% NP4EC), NP3EC (5.0% NP1EC, 5.7% NP2EC, 80.0% NP3EC, 0.0% NP4EC), and NP4EC (5.7% NP1EC, 3.1% NP2EC, 16.8% NP3EC, 60.0% NP4EC); mass spectra are for the first group 5 peak (denoted by *) in the NP1EC-NP4EC isomeric cluster]

of NP1EO contains diagnostic ions at *m/z* 165, 207, and 221 attributed to addition of one EO unit (*m/z* 44) to the ions corresponding to *m/z* 121, 163, and 177 in NP. Similar ions that are 44, 88, and 132 amu greater are observed for this isomer of NP2EO, NP3EO, and NP4EO, and the relative abundance of the various ions changes with increasing EO units. There is a progressive decrease in the relative abundance of [M^+ - 99] from the base peak in NP1EO (*m/z* 165) to non-existent in NP4EO (*m/z* 297). As was the case for NP, the NPEO spectra contain the ion series *m/z* 107+14*n*. Other features of the NPEO mass spectra are the ions at *m/z* 45 for NP1EO and *m/z* 45 and 89 for NP2EO-NP4EO, which are assigned to [CH_2CH_2OH] and [$CH_2CH_2OCH_2CH_2OH$]. The intensity of the *m/z* 45 ion increases with increasing EO length and eventually becomes the base peak. As was the case for OP, the OPEO homologs (data not shown) occur as single isomers with mass spectra similar to the α,α-dimethyl configuration of group 1 NP, but with [M^+] at *m/z* 250, 294, 338, and 382 for OP1EO-OP4EO respectively.

The methyl esters of NPEC elute as a series of homologs similar to NPEO, each an isomeric mixture (Figure 3a). The mass spectra of the methyl esters of NP1EC-NP4EC have [M^+] at *m/z* 292, 336, 380, and 424 (Figure 3b-e) representing addition of 1-4 ethoxycarboxylate units to NP. The NP1EC spectra (first eluting group 5 isomer) has a base peak at *m/z* 193, representing loss of both the propyl and butyl groups, and significant ions at *m/z* 235 and 249, representing loss of either the propyl or the butyl group. The NP2EC and NP4EC spectra have a base peak at *m/z* 117, representing the methyl ester of the carboxylated 2EO side chain [$CH_2CH_2OCH_2CO_2CH_3$] which also is a significant ion in NP3EC. The ions at *m/z* 293 and 279 for NP2EC, *m/z* 337 and 323 for NP3EC, and *m/z* 381 and 367 for NP4EC represent [M^+ - C_3H_7] and [M^+ - C_4H_9]. The ion at *m/z* 59 represents the methylated carboxyl group [CO_2CH_3]. The NPEC propyl esters have similar chromatography and mass spectra as the methyl esters (data not shown) although the [M^+] and other ions are *m/z* 28 higher than the methyl esters due to the two additional methylene groups.

Full scan mass spectra were used to develop the SIM method (ions listed in Table II). Compound identity in the environmental samples was confirmed by matching full scan mass spectra and retention times against authentic standards. The sum of the peak areas of the three ions collected in SIM mode for each compound (isomer or homolog) was used for quantitation. Total NP, NPEO, and NPEC concentrations were determined by summing the area for the three ions across the retention time window for the entire isomer cluster of each homolog. An alternative method is to quantify the individual isomers and then sum the concentrations. Quantitation of the NPEO and OPEO compounds is problematic because of the assumptions required to define the homolog distributions of the technical mixtures. There currently are no high purity standards available for the NPEC compounds, which further confounds the quantitation.

Spike and Recovery Experiments. Spike and recovery experiments were carried out in distilled water and are summarized in Table II. Recoveries of the determined compounds ranged from 20-130% at a concentration of 0.5 μg/L, with relative standard deviations ranging from 9.9-53%. The lowest recoveries were for the antioxidant compounds (DTBP, BHT) which apparently oxidized to DTBB during extraction (*26*). Nonylphenol and NP1EO-NP4EO had recoveries from 60-80%. Recoveries were greater for the single isomer *n*-NP compounds that are easier to quantify than the isomeric mixtures. Recoveries for the NP1EC-NP4EC methyl esters ranged from 60-100%, and recovery for bisphenol A was 96%. Recoveries for the underivatized steroids were less than 50%, likely due to losses in the chromatographic system and sorption to the glassware. Minimum detection limits for NP, NP1EO, NP2EO, bisphenol A, and 17-β-estradiol, were 0.165 μg/L, 0.325 μg/L, 0.350 μg/L, 0.059 μg/L, and 0.058 μg/L respectively.

Field Results. There was considerable variation in the inorganic and organic composition of the different effluents and river waters as indicated by concentrations of ammonium, TOC, and EDTA (Tables III and IV). Concentrations of TOC in the effluents ranged from 7.5-16.9 mg/L in October 1997 and from 6.3-16.7 mg/L in February 1998. During both periods, the Jones Island and Detroit plants had the highest TOC concentration and the Northside plant had the lowest. Ammonium concentrations in the wastewater ranged from 0.11-11.9 mg/L in the fall and from 0.21-7.9 mg/L in the spring. Ammonium was highest in the Detroit plant for both samplings, and was lowest at the Northside plant in October and at the Calumet plant in February. During the fall sampling, EDTA concentrations in the wastewater ranged from 170-436 μg/L, with the highest values at Seneca and Blue Lake and the lowest values at Northside and Stickney. In the spring, concentrations of EDTA in the effluents ranged from 132-439 μg/L, with highest concentrations at Seneca and Blue Lake and lowest concentrations at Detroit and Northside.

During October 1997, TOC concentrations in the three rivers ranged from 6.0-8.5 mg/L. The concentration of TOC in the Des Plaines River (6.0 mg/L), more than 30 miles downstream of the three Chicago wastewater discharge points (Calumet, Northside, and Stickney), was 30% less than the average concentration for the Chicago effluents (8.6 mg/L). There was no change in TOC between the Des Plaines River and Illinois River sites, an additional 55 miles downstream. In contrast, EDTA concentrations decreased 59% at the Des Plaines River site and 73% at the Illinois River site relative to the average wastewater value (187 μg/L). The Minnesota River site is upstream from the three Minneapolis/St. Paul treatment plant discharges (Metropolitan, Seneca, and Blue Lake) but receives discharges from cities further upstream. The concentrations of TOC and EDTA in the Minnesota River were less than in the three treatment plant effluents. Although the concentration of EDTA in the Minnesota River was less than in the Des Plaines and Illinois Rivers, the presence of EDTA indicates impact by wastewater disposal. Ammonium concentrations in the Des Plaines River were 46% of the average for the three Chicago treatment plants, and the concentration decreased 88% between the Des Plaines River and Illinois River sites. The concentration of ammonium in the Minnesota River was similar to the Des Plaines River and was less than for treated wastewater.

In the spring 1998, concentrations of TOC and ammonium in the Des Plaines River were only slightly less than the average for the three Chicago effluents. There was little change in TOC concentrations between the Des Plaines River and Illinois River sites, but ammonium concentrations decreased by 77%.

Table III. Results for CLLE analysis of samples collected during October 1997. Site abbreviations are listed in Table I. [concentrations in µg/L unless indicated otherwise; MGD, million gallons per day; na, not available; nd, not detected at ~0.01 µg/L; %, percent recovery for surrogate standards based on concentrations calculated from external calibration curves]

COMPOUND	IR	DR	MR	ST	NS	CA	SE	ME	BL	JI	DE
Sampling date	10/8	10/8	10/13	10/9	10/9	10/9	10/14	10/13	10/13	10/8	10/8
Flow (MGD)	2,778[a]	2,077[b]	1,020[c]	975	270	na	24.2	236	31.8	115	na
TOC (mg/L)	6.0	6.0	8.5	8.8	7.5	9.6	9.8	10.2	10.3	16.9	15.2
EDTA	51	76	31	170	172	219	436	219	390	219	242
ammonium (mg/L)	0.05	0.41	0.42	1.38	0.11	0.83	2.42	0.20	2.97	0.15	11.9
4-methylphenol	0.03	0.03	nd	0.04	0.04	0.05	0.06	0.04	0.07	0.03	1.1
4-*t*-butylphenol	nd	nd	nd	nd	nd	nd	nd	nd	nd	nd	1.8
DTBP	nd	nd	nd	nd	nd	0.06	nd	nd	nd	nd	nd
DTBB	nd	nd	nd	0.31	0.20	0.45	0.43	0.46	0.35	0.26	1.8
BHA	nd	nd	nd	0.07	0.04	nd	0.07	0.04	0.26	0.09	nd
BHT	nd	nd	nd	0.05	0.05	0.12	0.11	0.10	0.21	0.05	0.23
OP	nd	nd	nd	nd	nd	nd	nd	nd	0.14	nd	1.3
NP	nd	1.7	nd	2.1	1.2	2.2	1.7	2.3	2.2	2.1	23
NP1EO	nd	0.40	nd	2.7	1.7	1.7	2.7	2.5	1.6	5.0	32
NP2EO	nd	0.26	nd	2.4	1.1	1.0	3.0	2.6	0.84	8.0	47
NP3EO	nd	nd	nd	nd	nd	nd	nd	nd	nd	1.0	15
NP4EO	nd	nd	nd	nd	nd	nd	nd	nd	nd	nd	2.0
caffeine	0.13	0.23	0.09	0.20	0.12	0.18	0.15	0.20	0.22	0.13	3.7
bisphenol A	nd	0.06	nd	nd	nd	0.16	nd	nd	nd	nd	nd
coprostanol	nd	nd	nd	0.60	0.30	0.40	0.50	0.40	0.31	0.30	3.5
cholesterol	nd	0.45	0.90	1.5	1.0	0.90	1.1	1.2	1.3	2.6	6.7
NP1EC	0.10	2.6	0.80	11	7.9	7.9	7.5	4.3	19	3.8	140
NP2EC	2.9	7.9	1.1	45	21	25	33	21	16	44	100
NP3EC	0.60	1.9	nd	6.5	2.4	7.2	14	8.7	3.6	9.7	12
NP4EC	0.10	0.10	0.10	1.5	0.8	0.60	1.8	1.9	0.50	3.0	3.4
Surrogate Standards	%	%	%	%	%	%	%	%	%	%	%
d_{21}-BHT	70	102	59	92	102	122	88	93	122	106	64
d_6 bisphenol A	82	82	105	127	76	93	120	94	131	128	90

[a] Provisional USGS stream flow data in cubic feet per second (CFS) at river mile 247, Marseilles, IL (Amanda Kupka, USGS).

[b] Sum of provisional USGS stream flow data in CFS from the Des Plaines River at Riverside, IL and the Chicago Sanitary and Ship Canal at Romeoville, IL (Amanda Kupka, USGS).

[c] Provisional USGS stream flow data in CFS from river mile 39.6 (Ron Jacobson, MPCA).

Table IV. Results for analysis of CLLE samples collected during February and March 1998. Site abbreviations are listed in Table I. [concentrations in µg/L unless indicated otherwise; MGD, million gallons per day; nd, not detected at ~0.01 µg/L; %, percent recovery for surrogate standards based on concentrations calculated from external calibration curves]

COMPOUND	IR	DR	MR	ST	NS	CA	SN	ME	BL	JI	DE
Sampling date	3/4	3/4	2/24	2/25	2/25	2/25	2/24	2/25	2/24	2/24	2/24
Flow (MGD)	6,279[a]	1,930[b]	4,974[c]	855	114	235	24.7	207	23.7	96.2	946
TOC (mg/L)	7.0	6.9	10.4	7.6	6.3	7.6	8.5	11.6	9.6	16.7	12.2
EDTA	19	87	3.6	175	137	170	439	336	383	175	132
ammonium (mg/L)	0.14	0.60	0.67	0.56	1.0	0.21	4.6	4.0	0.31	1.9	7.9
4-methylphenol	nd	nd	nd	nd	nd	nd	0.02	nd	nd	nd	0.80
4-*t*-butylphenol	nd	nd	nd	0.05	0.15	nd	0.28	0.25	0.06	0.25	1.0
4-*t*-pentylphenol	nd	nd	nd	0.30	nd	nd	nd	0.27	nd	0.25	0.90
DTBP	nd	nd	nd	nd	nd	nd	nd	0.17	0.13	nd	0.14
DTBB	nd	0.20	nd	0.47	0.64	0.56	0.90	0.80	0.81	0.70	1.4
BHA	0.17	0.25	nd	0.58	0.61	0.64	1.3	1.3	0.89	0.16	0.17
BHT	nd	nd	nd	0.03	0.05	0.08	0.08	nd	0.30	0.03	0.09
n-OP	nd	nd	nd	nd	nd	nd	nd	0.20	nd	nd	nd
OP	nd	nd	nd	nd	nd	nd	nd	nd	nd	nd	1.4
NP	nd	1.1	nd	1.7	1.6	1.4	1.5	2.1	0.90	2.5	19
NP1EO	nd	1.6	nd	4.1	13	5.8	6.0	12	1.5	9.0	55
NP2EO	nd	0.50	nd	5.5	7.0	3.7	2.9	19	0.78	29	110
NP3EO	nd	nd	nd	nd	nd	nd	nd	2.0	nd	3.0	17
NP4EO	nd	nd	nd	nd	nd	nd	nd	1.2	nd	nd	5.0
OP1EO	nd	nd	nd	0.50	1.1	0.50	0.65	1.2	0.25	0.84	3.2
OP2EO	nd	nd	nd	1.5	1.5	0.80	0.80	6.5	0.78	6.0	25
OP3EO	nd	nd	nd	nd	nd	nd	nd	4.6	nd	nd	23
caffeine	0.17	0.22	0.10	0.14	0.14	0.34	0.13	2.4	0.15	0.77	4.0
bisphenol A	nd	0.20	nd	0.38	nd	0.55	nd	nd	nd	0.94	2.7
coprostanol	nd	0.10	nd	0.47	0.60	0.31	0.15	1.3	0.14	0.80	14
cholesterol	0.60	0.60	0.60	1.2	1.1	1.0	0.70	2.2	0.61	1.6	8.0
NP1EC	3.3	19	1.2	16	26	29	60	25	21	48	34
NP2EC	5.3	31	0.60	92	64	44	80	56	100	120	70
NP3EC	0.40	2.7	0.20	9.5	4.1	4.3	7.8	13	11	16	4.6
NP4EC	0.30	0.70	0.10	6.1	1.3	1.7	2.3	6.5	5.4	3.6	1.8
Surrogate Standards	%	%	%	%	%	%	%	%	%	%	%
d_{21}-BHT	55	47	57	38	70	91	36	43	70	21	65
d_6 bisphenol A	70	94	111	108	110	114	104	122	114	94	120
n-NP	71	91	104	82	120	125	95	134	131	96	123
17-α-estradiol	84	82	92	99	75	105	80	109	111	83	126
$^{13}C_6$ *n*-NP	91	80	149	90	122	124	108	130	137	107	131
$^{13}C_6$ *n*-NP1EO	80	129	126	137	152	137	155	206	142	210	257
$^{13}C_6$ *n*-NP2EO	35	43	51	51	48	41	74	113	46	109	117
$^{13}C_6$ *n*-NP3EO	nd	nd	nd	nd	nd	nd	nd	nd	nd	nd	nd
BPAA[d]	33	29	26	26	29	36	20	50	27	31	18

[a] Provisional USGS stream flow data in cubic feet per second (CFS) at river mile 247, Marseilles, IL (Amanda Kupka, USGS).

[b] Sum of provisional USGS stream flow data in CFS from the Des Plaines River at Riverside, IL and the Chicago Sanitary and Ship Canal at Romeoville, IL (Amanda Kupka, USGS).

[c] Provisional USGS stream flow data in CFS from river mile 39.6 (Ron Jacobson, MPCA).

[d] BPAA absolute recoveries. Concentrations of NPEC are calculated using the surrogate standard method which collects for recovery.

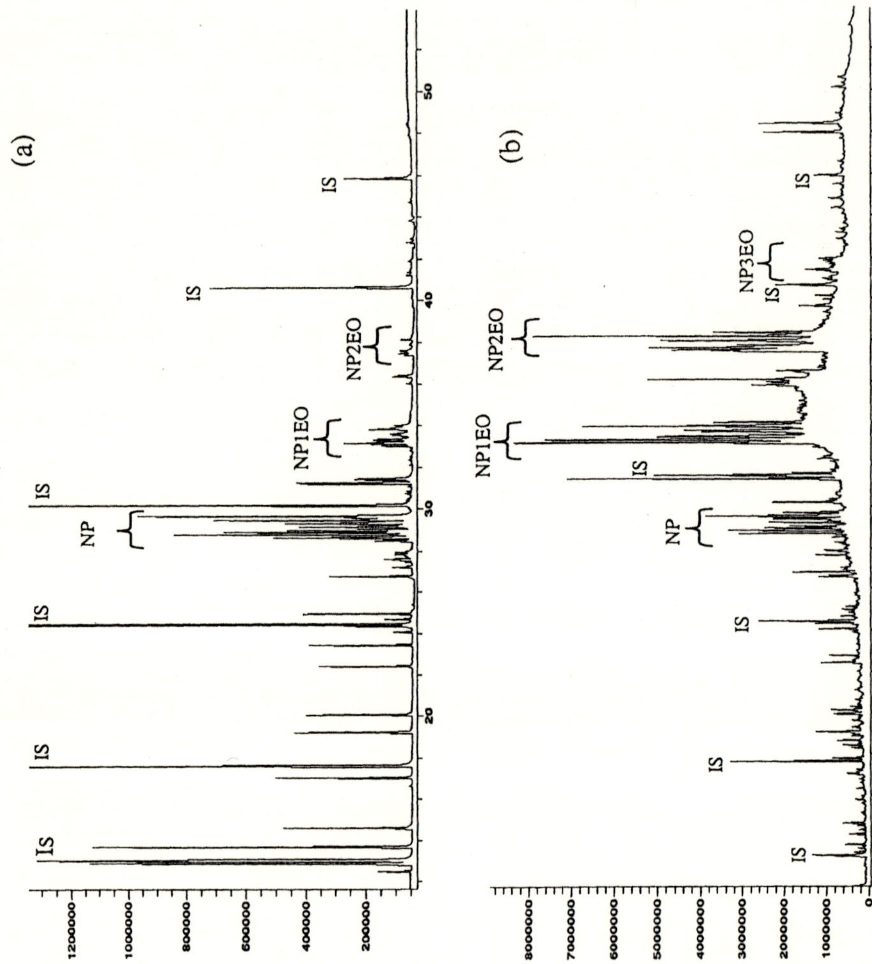

(a)
IS
IS
IS
NP
IS
NP1EO
NP2EO
IS
IS
Abundance
1200000
1000000
800000
600000
400000
200000
20
30
40
50
(b)
IS
IS
IS
NP
IS
NP1EO
NP2EO
IS
NP3EO
IS
Abundance
8000000
7000000
6000000
5000000
4000000
3000000
2000000
1000000
0
20
30
40
50

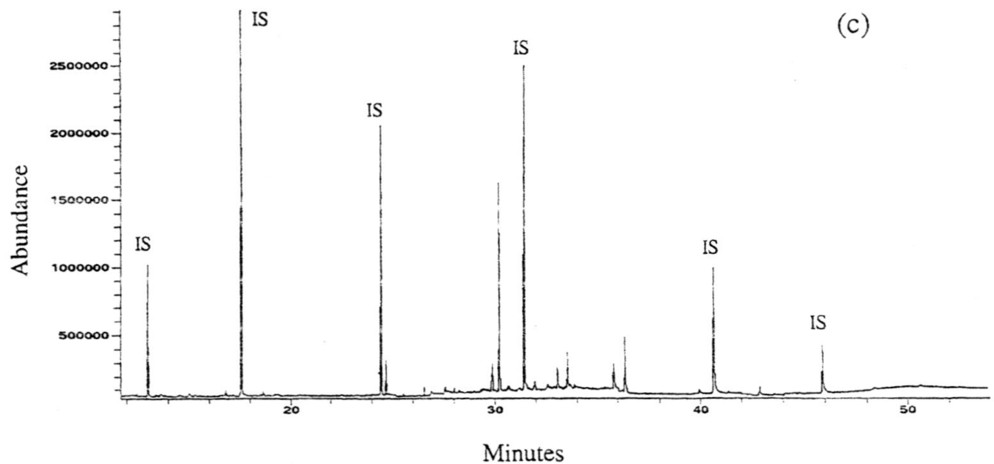

Figure 4. Total ion chromatograms from SIM GC/MS analysis of CLLE extracts for (a) 1 μg/L standard mix, (b) Detroit wastewater, February 1998, and (c) Minnesota River, February 1998. [IS, internal standard; NP, 4-nonylphenol; NP1EO-NP3EO, 4-nonylphenol mono-, di-, and triethoxylate]

The concentration of EDTA in the Des Plaines River was similar to the fall value, and there was a 78% decrease between the Des Plaines and Illinois River sites. The concentration of EDTA in the Minnesota River was significantly less than in the fall sampling, but TOC and ammonium concentrations were higher.

The chromatography of compounds determined by the CLLE GC/MS method was complex (Figure 4) because NP, NPEO, and OPEO are mixtures of isomers and homologs. In the fall 1997 samples (Table III) 4-methylphenol, BHT, DTBB, NP, NP1EO, NP2EO, caffeine, coprostanol, cholesterol, and NP1EC-NP4EC were detected in all of the effluents. Concentrations of NP ranged from 1.2-23 μg/L, the sum of NP1EO-NP2EO ranged from 2.4-79 μg/L, the sum of NP1EC-NP2EC ranged from 25-240 μg/L, coprostanol ranged from 0.30-3.5 μg/L, and cholesterol ranged from 0.90-6.7 μg/L. Bisphenol A was detected in one wastewater effluent and one river sample at 0.16 μg/L and 0.06 μg/L respectively. None of the samples had detectable concentrations of 17-β-estradiol, estrone, or 17-α-ethynylestradiol. Samples were analyzed for 17-β-estradiol by radioimmunoassay and all concentrations were below the detection limit (~0.02 μg/L). All of the river samples had detectable concentrations of NPEC and caffeine. The Des Plaines River site had detectable concentrations of NP, NP1EO-NP2EO, and bisphenol A, whereas these compounds were below detection further downstream at the Illinois River site. Concentrations of total NPEC decreased 73% at the Des Plaines River site and 92% at the Illinois River site relative to the average for the Chicago effluents (45.6 μg/L), a greater decrease than for EDTA. The Minnesota River had detectable concentrations of caffeine, cholesterol, and NPEC, which were lower than in the Des Plaines and Illinois Rivers.

In the spring 1998 samples (Table IV) DTBB, BHA, NP, NP1EO, NP2EO, OP1EO (not analyzed in October), caffeine, coprostanol, cholesterol, and NP1EC-NP4EC were detected in all of the effluents, and BHT was detected in all but one of the effluents. Concentrations of NP ranged from 0.9-19 μg/L, NP1EO-NP2EO ranged from 2.3-165 μg/L, and NP1EC-NP2EC ranged from 73-168 μg/L, similar to values observed in the fall. During both samplings, the relative amounts of NP1EO and NP2EO were about the same, and NP3EO and NP4EO were only minor components. In contrast, NP2EC was always more abundant than NP1EC, but NP3EC and NP4EC were only minor components. Results for analysis of NPEC as both the methyl and propyl esters were similar for both sampling rounds (data not shown), with respect to both concentrations and homolog distributions. The total concentration of APEO derived compounds ranged from 93-370 μg/L, which was consistent with the fall samples (36-380 μg/L). All of the effluent samples had coprostanol (0.14-14 μg/L) and cholesterol (0.60-8.0 μg/L). Bisphenol A was detected in half of the effluents at concentrations ranging from 0.38-2.7 μg/L. None of the samples had detectable 17-β-estradiol, estrone, or 17-α-ethynylestradiol. Samples were analyzed for 17-β-estradiol by radioimmunoassay and all were below the detection limit. All of the river samples had detectable NPEC, cholesterol, and caffeine. As was observed in the fall, the Des Plaines River site had bisphenol A, NP, NP1EO, and NP2EO, whereas these compounds were below the detection limit at the Illinois River site. Concentrations of total NPEC decreased 46% at the Des Plaines River site and 91% at the Illinois River site relative to the average for the Chicago effluents (99.3 μg/L). Caffeine, coprostanol, and NPEC were detected in the Minnesota River at concentrations lower than in the Des Plaines and Illinois Rivers.

Discussion

The wastewater effluents analyzed in this study contained a complex mixture of AP, APEO, and APEC at concentrations ranging from <0.01-140 μg/L for individual compounds and homologs, and from 36-380 μg/L for total APEO compounds. Concentrations of NP ranged from 1.2-23 μg/L in the fall and from 0.9-19 μg/L in the spring, and represent total NP (also the case for NPEO and NPEC) which includes dissolved, particulate, and colloidal phases. The treated wastewater effluents and river waters analyzed in this study (with the exception of the Minnesota River) had low suspended solids, and it is likely that the NPEO compounds were primarily dissolved as was observed by Naylor and others (*34*). Ahel and others (*14*) reported that up to 90% of the NP and NPEO in a wastewater treatment plant is associated with solid particles, whereas less than 10% of the NPEC is particle bound. The concentrations of NP1EO-NP2EO measured in this study are lower than values reported for wastewater in Switzerland in 1983 (*14*), and are comparable to present day Swiss effluent concentrations (*33*). Studies on the Fox River of Wisconsin (*34, 45*) report that treated wastewater had concentrations of NP, NPEO, and NPEC that are, for the most part, in agreement with those measured here. Likewise the concentrations of NP, OP, and NP1EO-NP2EO reported here are consistent with concentrations measured in effluents and surface waters in Canada (*36, 37*).

Several basic water quality parameters were measured in addition to specific EDC. These parameters include ammonium, TOC, and EDTA. There was a general positive correlation between treatment efficiency, as indicated by ammonium and TOC concentrations, and the concentrations of NPEO derived compounds. For example, the Detroit effluent had high concentrations of ammonium and TOC, and also had high concentrations of NP, OP, NPEO, OPEO, and NPEC. However, the concentrations of NPEO derived compounds in treated effluent also reflect the composition of the sewage influent to the treatment plant, which is a function of the relative proportions and chemical consumption patterns of domestic, commercial, and industrial inputs. There also was a relationship between the level of treatment and the composition of the NPEO compounds. The Detroit effluent had greater concentrations of NPEO than NPEC, in contrast to the other plants that had higher concentrations of NPEC than NPEO. Wastewater with more NPEC than NPEO indicates aerobic conditions, whereas wastewater with more NPEO than NPEC indicates anaerobic conditions (*14*), which is consistent with the types of treatment at the various plants (Table I). In contrast to results for the NPEO derived compounds, concentrations of EDTA, which is less susceptible to removal by sorption and biodegradation in the treatment process, did not show a strong correlation to TOC and ammonium. The EDTA concentrations in the effluents most likely reflected varying concentrations in the sewage influents, which is a function of the relative proportions and types of industries contributing inflow to the treatment plants.

Once wastewater is discharged into a river environment, a number of physical, chemical, and biological processes act to decrease concentrations of organic contaminants. The primary physical process influencing concentrations of contaminants is dilution of wastewater with river water as reflected by stream dilution factors, the ratio between minimum river flow (7 day low flow for a 10 year period) and treatment plant design flow volume (Table I). The systems investigated in this study had minimum stream dilution factors ranging from <0.001 (Northside) to 14 (Detroit). Large volume rivers such as the Detroit River have greater assimilation capacity from dilution then do wastewater-dominated streams such as the Des Plaines River. Thus, even though the Detroit effluent had greater concentrations of APEO compounds, concentrations in the Detroit River were likely lower than in the Des Plaines River which had a higher proportion of wastewater. Total organic carbon in rivers consists of natural organic matter such as humic substances, which are recalcitrant to biological and chemical degradation, as well as wastewater derived organic contaminants, which can be more degradable. The results for both samplings showed little loss of TOC between the Des Plaines River and Illinois River sites. In contrast, there was a significant decrease in the concentration of EDTA between these two sites (33% in the fall and 78% in the spring), indicating instream removal or dilution by waters with similar TOC concentrations. The decrease in concentrations of NPEC was greater than for EDTA (70% in the fall and 83% in the spring), although both compounds persist for considerable distance in the river. Most of the other compounds detected at the Des Plaines River site appeared to be removed by instream processes and were below detection limits at the Illinois River site.

The widespread occurrence of APEO derived compounds in treated wastewater raises concerns about their impact on the environment following disposal of effluent into aquatic systems. Jobling and others (*22*) report a million-fold increase in plasma levels of the egg yoke precursor vitellogenin (VTG), and a 50% inhibition of testicle growth in male rainbow trout exposed to 30 μg/L of OP. Exposure to similar concentrations of NP, NP2EO, and NP1EC increased VTG concentrations by 100-1000 fold. The synthetic compound 17-α-ethynylestradiol caused a similar effect at 0.002 μg/L as 30 μg/L of OP. The lowest concentration of NP required to induce production of VTG was 20 μg/L, whereas OP induced VTG at 4.8 μg/L. The relative estrogenic potencies were OP>NP=NP2EO=NP1EC, which is consistent with earlier observations from *in vitro* studies (*20*). The concentrations required to induce responses in *in vivo* tests was two orders of magnitude lower than in *in vitro* tests due to bioaccumulation effects or metabolism by the fish into more active metabolites. Concentrations of NP in the effluents reported here were a factor of 2-10 below the no observable effect concentrations (NOEC), and concentrations of OP were about a factor of 4 less. However, at some sites concentrations of NP1EO-NP2EO and NP1EC-NP2EC were above values demonstrated to induce VTG production.

Folmar and others (*6*) collected carp from near the effluent discharge of the Metropolitan Plant in Minneapolis/St. Paul, MN, and reported indications of endocrine disruption in the form of elevated plasma VTG and decreased plasma testosterone levels in male carp relative to control fish. Carp collected from the Minnesota River showed decreased plasma testosterone levels but no serum VTG.

The results of our study indicate that based on NOEC reported in the literature, there may be sufficient concentrations of total AP, APEO, and APEC in the Metropolitan Plant effluent (43-150 μg/L) to cause the observed responses, whereas the Minnesota River had significantly lower levels (2 μg/L).

Summary

An analytical method was developed for the measurement of potential EDC in treated wastewater and natural waters. The method involves isolation of the compounds from water by CLLE, derivatization of the polar and ionic compounds, and analysis of the various fractions by GC/MS in full scan and SIM modes. Analyses of a variety of treated municipal wastewaters indicate that concentrations of individual AP compounds are below levels likely to cause endocrine disrupting effects in fish. However, the total APEO concentrations are relatively high and may affect organisms living near effluent disposal sites and in low dilution receiving streams. Although the detection limit for the method using a 1-liter sample was not adequate to detect endogenous and synthetic sex hormones in the effluent samples, these compounds may still have estrogenic effects on fish because of their extreme potency.

Acknowledgements

We thank Colleen Rostad (US Geological Survey), Jeff Writer (Boulder Valley School District), Adrian Pike and David Norris (University of Colorado), Peter Howe, Al Alwan, William Sargent, John Dorkin, John McGuire, and Larry Lins (US Environmental Protection Agency), George Richardson (Metropolitan Water Reclamation District of Greater Chicago), John Moser (Milwaukee Metropolitan Sanitation District), Kent Johnson (Minneapolis Metropolitan Council), and Steve Kuplicki (City of Detroit Water and Sewage Department) for their assistance with this study. Use of trade names in this paper is for identification purposes and does not imply endorsement by the USGS.

Literature Cited

1. Colborn, T.; vom Saal, F.S.; Soto, A.M. *Environ. Health Perspect.* **1993**, *v. 101*, pp. 378-384.

2. National Institute of Health *Environ. Health Perspect.* **1995**, *v. 103*, *Suppl. 7*, 184 p.

3. Colborn, T.; Dumanoski, D.; Myers, J.P. *Our Stolen Future*; Plume/Penguin: New York, NY, **1996**; 316 p.

4. U.S. Environmental Protection Agency *EPA/630/R-96/012* **1997**, 111 p.

5. Purdom, C.E.; Hardiman, P.A.; Bye, V.J.; Eno, N.C.; Tyler, C.R.; Sumpter, J.P. *Chem. Ecol.* **1994**, *v. 8*, pp. 275-285.

6. Folmar, L.C.; Denslow, N.D.; Rao, V.; Chow, M.; Crain, D.A.; Enblom, J.; Marcino, J.; Guillette, J.J., Jr. *Environ. Health Perspect.* **1996**, *v. 104*, pp. 1096-1101.

7. Harries, J.E.; Sheahan, D.A.; Jobling, S.; Matthiessen, P.; Neall, P.; Routledge, E.J.; Rycroft, R.; Sumpter, J.P.; Tylor, T. *Environ. Toxicol. Chem.* **1996**, *v. 15*, pp. 1993-2002.

8. Desbrow, C.; Routledge, E.J.; Brighty, G.C.; Sumpter, J.P.; Waldock, M. *Environ. Sci. Technol.* **1998**, *v. 32*, pp. 1549-1558.

9. Jobling, S.; Nolan, M.; Tyler, C.R.; Brighty, G.; Sumpter, J.P. *Environ. Sci. Technol.* **1998**, *v. 32*, pp. 2498-2506.

10. Routledge, E.J.; Sheahan, D.; Desbrow, C.; Brighty, G.C.; Waldock, M.; Sumpter, J.P. *Environ. Sci. Technol.* **1998**, *v. 32*, pp. 1559-1565.

11. Talmage, S.S. *Environmental and Human Safety of Major Surfactants–Alcohol Ethoxylates and Alkylphenol Ethoxylates*; Lewis Publishers: Boca Raton, FL, **1994**; 374 p.

12. Swisher, R.D. *Surfactant Biodegradation, Second Ed.*; Marcel Dekker, Inc.: New York, NY, **1987**; 1085 p.

13. Ahel, M.; Giger, W.; Koch, M. *Water Res.* **1994**, *v. 28*, pp. 1131-1142.

14. Ahel, M.; Giger, W.; Schaffner, C. *Water Res.* **1994**, *v. 28*, pp. 1143-1152.

15. Giger, W.; Brunner, P.H.; Schaffner, C. *Sci.* **1984**, *v. 225*, pp. 623-625.

16. McLeese, D.W.; Zitko, V.; Sargent, D.B.; Burridge, L.; Metcalfe, C.D. *Chemosphere* **1981**, *v. 10*, pp. 723-730.

17. Thiele, B.; Gunther, K.; Schwuger, M.J. *Chem. Rev.* **1997**, *v. 97*, pp. 3247-3272.

18. U.S. Environmental Protection Agency *Fed. Reg.* **1996**, *v. 61*, pp. 4188-4196.

19. Soto, A.M.; Justicia, H.; Wray, J.W.; Sonnenschein, C. *Environ. Health Perspect.* **1991**, *v. 92*, pp. 167-173.

20. Jobling, S.; Sumpter, J.P. *Aquatic Toxicol.* **1993**, *v. 27*, pp. 361-372.

21. White, R.; Jobling, S.; Hoare, S.A.; Sumpter, J.P.; Parker, M.G. *Endocrinology* **1994**, *v. 135*, pp. 175-182.

22. Jobling, S.; Sheahan, D.; Osborne, J.A.; Matthiessen, P.; Sumpter, J.P. *Environ. Toxicol. Chem.* **1996**, *v. 15*, pp. 194-202.

23. Routledge, E.J.; Sumpter, J.P. *Environ. Toxicol. Chem.* **1996**, *v. 15*, pp. 241-248.

24. Routledge, E.J.; Sumpter, J.P. *J. Biol. Chem.* **1997**, *v. 272*, pp. 3280-3288.

25. Kirk-Othmer *Encyclopedia of Chemical Technology, 4th Edition*; John Wiley and Sons: New York, NY, **1992**.

26. Lopez-Avila, V.; Hites, R.A. *Environ. Sci. Technol.* **1981**, *v. 15*, pp. 1386-1388.

27. Tomson, M.B.; Dauchy, J.; Hutchins, S.; Curran, C.; Cook, C.J.; Ward, C.H. *Water Res.* **1981**, *v. 15*, pp. 1109-1116.

28. Rudel, R.A.; Melly, S.J.; Geno, P.W.; Sun, G.; Brody, J.G. *Environ. Sci. Technol.* **1998**, *v. 32*, pp. 861-869.

29. Barber, L.B., II; Thurman, E.M.; Schroeder, M.P.; LeBlanc, D.R. *Environ. Sci. Technol.* **1988**, *v. 22*, pp. 205-211.

30. Jungclaus, G.A.; Lopez-Avila, V.; Hites, R.A. *Environ. Sci. Technol.* **1978**, *v. 12*, pp. 88-96.

31. Giger, W.; Stephanou, E.; Schaffner, C. *Chemosphere* **1981**, *v. 10*, pp. 1253-1263.

32. Stephanou, E.; Giger, W. *Environ. Sci. Technol.* **1982**, *v. 16*, pp. 800-805.

33. Ahel, M; Giger, W. *Am. Chem. Soc., Nat. Meeting Extended Abst.* **1998**, *v. 38*, no. 2, pp. 276-278.

34. Naylor, C.G.; Williams, J.B.; Webb, D.A. *The Alkylphenols and Alkylphenol Ethoxylates Review* **1998**, pp. 44-53.

35. Paxeus, N.; Robinson, P.; Balmer, P. *Water Sci. Technol.* **1992**, *v. 25*, pp. 249-256.

36. Lee, H.B.; Peart, T.E. *Anal. Chem.* **1995**, *v. 67*, pp. 1976-1980.

37. Bennie, D.T.; Sullivan, C.A.; Lee, H.B.; Peart, T.E.; Maguire, R.J. *Sci. Total Environ.* **1997**, *v. 193*, pp. 263-275.

38. Ahel, M.; Giger, W. *Anal. Chem.* **1985**, *v. 57*, pp. 1577-1583.

39. Marcomini, A.; Giger, W. *Anal. Chem.* **1987**, *v. 59*, pp. 1709-1715.

40. Kubeck, E.; Naylor, C.G. *J. Am. Oil Chem. Soc.* **1990**, *v. 67*, pp. 400-404.

41. Naylor, C.G.; Mieure, J.P.; Adams, W.J.; Weeks, J.A.; Castaldi, F.J.; Ogle, L.D.; Romano, R.P. *J. Am. Oil Chem. Soc.* **1992**, *v. 69*, pp. 695-703.

42. Di Corcia, A.; Samperi, R.; Marcomini, A. *Environ. Sci. Technol.* **1994**, *v. 28*, pp. 850-858.

43. Ahel, M.; Conrad, T.; Giger, W. *Environ. Sci. Technol.* **1987**, *v. 21*, pp. 697-703.

44. Clark, L.B.; Rosen, R.T.; Hartman, T.G.; Louis, J.B.; Suffet, I.H.; Lippincott, R.L.; Rosen, J.D. *Inter. J. Environ. Anal. Chem.* **1992**, *v. 47*, pp. 167-180.

45. Field, J.F.; Reed, R.L. *Environ. Sci. Technol.* **1996**, *v. 30*, pp. 3544-3550.

46. Fujita, Y.; Ding, W.H.; Reinhard, M. *Water Environ. Res.* **1996**, *v. 68*, pp. 867-876.

47. Schaffner, C.; Giger, W. *J. Chrom.* **1984**, *v. 312*, pp. 413-421.

48. Tabak, H.H.; Bloomhuff, R.N.; Bunch, R.L. *Dev. Ind. Microbiol.* **1981**, *v. 22*, pp. 497-519.

49. Staples, C.A.; Dorn, P.B.; Klecka, G.M.; O'Block, S.T.; Harris, L.R. *Chemosphere* **1998**, *v. 36*, pp. 2149-2173.

50. Nagel, S.C.; vom Saal, F.S.; Thayer, K.A.; Dhar, M.G.; Boechler, M.; Welshons, W.V. *Environ. Health Perspect.* **1997**, *v. 105*, pp. 70-76.

51. Thenot, J.P.; Horning, E.C. *Anal. Lett.* **1972**, *v. 5*, pp. 21-33.

52. Sjovall, J.; Axelson, M. *Vitamins Hormones* **1982**, *v. 39*, pp. 31-144.

53. Pike, A.W.; Moynihan, C.; Kibler, S.; Marsh, P.G.; Fennessey, P.V. *J. Chrom.* **1984**, *v. 306*, pp. 39-50.

54. Gaskell, S.J.; Gould, V.J.; Leith, H.M. In *Mass Spectrometry in Biomedical Research*; Gaskell, S.J., Ed.; John Wiley and Sons: New York, NY, **1986**, pp. 347-361.

55. Bhatt, B.D.; Prasad, J.V.; Kalpana, G.; Ali, S. *J. Chrom. Sci.* **1992**, *v. 30*, pp. 203-210.

56. Wheeler, T.F.; Heim, J.R.; LaTorre, M.R.; Janes, A.B. *J. Chrom. Sci.* **1997**, *v. 35*, pp. 19-30.

Chapter 8

Detection of Pharmaceuticals Entering Boston Harbor

R. Siegener and R. F. Chen

Environmental, Coast and Ocean Sciences, University of Massachusetts at Boston, Boston, MA 02125

In 1996 alone, $86 billion worth of pharmaceuticals were produced in the United States for use in human and veterinary medicine (*1*). Unlike suspected endocrine disruptors such as polychlorinated biphenyls (PCB's), dioxins, and nonylphenols, pharmaceuticals are purposely designed to be biologically active. This biological activity means that some pharmaceuticals (such as oral contraceptives) have the potential to act on naturally occurring populations at environmental concentrations when these chemicals are released into natural systems. At this time, little is known about the ultimate fate and transport of pharmaceuticals once they enter marine systems. Since the aquatic environment in many urban and industrial areas may contain organisms already stressed by poor water quality (*2*), the impact pharmaceuticals could have on the health of these organisms may be significant. As the EPA does not currently regulate the domestic disposal of pharmaceuticals (*3*), in-home use may be a significant source of pharmaceuticals to municipal waste streams and subsequently to the marine environment.

This study was designed to detect the presence of pharmaceuticals in sewage effluent entering Boston Harbor. The initial stage of this study focused on caffeine as a representative pharmaceutical with known, widely distributed inputs. Measurements of caffeine were made in sewage influent and effluent, and Boston Harbor water samples. Next, sewage effluent was extracted and analyzed for 17α-ethynilestradiol, an oral contraceptive known to have effects on aquatic life at concentrations of 0.1 – 10 ng L^{-1} (*4,5*). Finally, in a search for other detectable pharmaceuticals, unknown compounds present in the extracts were tentatively identified (as pharmaceuticals where appropriate), which should lend to a better overall understanding of the distribution of these compounds in the marine environment.

Methods

All samples were collected in 4-liter amber bottles and stored at 4° C for transport back to the laboratory. Extractions were performed within 24 hours of sampling.

Separate extraction methodologies were used for analysis of caffeine and 17α-ethynilestradiol. Unfiltered samples for caffeine analysis were split into 2-liter aliquots and then serially liquid/liquid extracted with 100 mL, 50 mL, and 50 mL aliquots of dichloromethane (DCM). In cases where emulsions were present, additional aliquots of DCM were used to extract the sample, or the emulsion itself was filtered through baked Na_2SO_4 to remove residual water. The extracts were then combined and rotary evaporated to ~10 mL. The 10 mL extracts were then reduced to ~100 μL under nitrogen and spiked with phenanthrene-d10 as an internal standard.

For 17α-ethynilestradiol analysis, difficulties with the extraction method prevented sample extraction within 24 hours of collection. Therefore, the samples were poisoned with mercuric chloride and refrigerated until extraction. Two 4 L effluent samples were pressure filtered through baked GFF filters using high purity nitrogen at 15 p.s.i. to remove the larger particles that tended to clog the Solid Phase Extraction (SPE) disks. Prior to extraction, the effluent was subsequently centrifuged at 4000 rpm for 30 minutes to remove the smaller particles still present in the effluent. 47 mm Empore (3M) C18 SPE disks were conditioned with 10 mL of methanol and then 10 mL of Milli-Q water as recommended by the manufacturer. A total of 7.08 L of effluent, in 500 – 650 mL aliquots, was drawn through the disks under vacuum. The disks (13 total) were then serially extracted three times with 10 mL aliquots of methanol. Extracts from individual disks were filtered through baked Na_2SO_4 and combined. The combined extract was then rotary evaporated to dryness and reconstituted with 10 mL of methanol. This 10 mL extract was then centrifuged at 1000 rpm for 15 minutes to remove a precipitate that formed during extraction. The supernatant was collected and then concentrated under nitrogen to a final volume of 245 μL.

Gas chromatography - mass spectrometry (GCMS) analysis was conducted using a Finnigan Voyager Mass Spectrometer system coupled to a Carlo Erba 8000 gas chromatograph. The gas chromatograph was equipped with a J&W DB-5ms column (30m x 0.25mm i.d. x 0.25 μm film). Helium (ultra-high purity) was used as the carrier gas with a flow rate of 1.0 mL $min.^{-1}$. Sample analysis for caffeine was performed using 1.0 μL splitless injections (split valve closed for 1 min.) and the following GC oven program: the oven was held at an initial temperature of 40° C for 1 minute, ramped at 10° C $min.^{-1}$, and held at 300° C for 5minutes. The injection port was heated to 275° C. Mass spectrometry was performed using electron impact (EI) ionization with the electron energy set to 70 eV. Under full scan conditions, the mass range from 50 to 450 Da was scanned with a scan rate of 2 scans s^{-1}.

Sample analysis for 17α-ethynilestradiol was performed using the same GC column and carrier gas flow conditions. For full scan analysis, 1.0 μL of sample was injected under splitless conditions into a 235° C injection port. A split time of 1 minute was used. The following GC program was used for both selected ion monitoring (SIM) and full scan analyses: the oven was held at 40° C for 1 minute, ramped at 2° C min^{-1} to a temperature of 300° C, then held at 300° C for 10 minutes. The mass spectrometer conditions for full scan analysis of 17α-ethynilestradiol were the same as those for caffeine. SIM analysis was performed under the same GC conditions, except a 2.0 μL injection was used. Selected ion monitoring (SIM) was performed on the effluent samples to detect ethynilestradiol. The SIM method scanned the following ions to detect ethynilestradiol: m/z = 91, 145, 157, 159, 161, 213, 214, and 296.

Results and Discussion

Caffeine. While caffeine is not an endocrine disruptor, its widespread use made it a logical choice as an initial representative pharmaceutical in natural waters. Currently the most widely used drug in the world, caffeine has been used as far back as the Paleolithic period (*6,7*). A highly water soluble compound, caffeine is present in a number of dietary products, such as coffee, tea, soft drinks, and chocolate, cold medications, analgesics, diuretics, and stimulants (*6-8*). The caffeine content of a 150 mL (5 oz.) serving of coffee ranges from 29 – 176 mg (*6*). Soft drinks typically contain anywhere from 6.2 – 62 mg of caffeine per 355 mL (12 oz.) (*9*). Daily consumption of caffeine from all sources (with beverages being the dominant source) by adults ranges from 200 – 280 mg per capita (*7,8*). In the calculations below, an estimated daily caffeine consumption of 206 mg per capita (*10*) was used, as this amount best represented consumption by the overall population.

Once ingested, caffeine from beverages is 99% absorbed from the gastrointestinal tract, and peak blood plasma levels are seen within 15 – 60 min. after ingestion (*11-13*). Caffeine is able to pass through all biological membranes and is therefore evenly distributed throughout the body (*11*). Metabolism takes place in the liver, with the primary metabolites being 1-methyl uric acid, 1-methyl xanthine, and 6-amino-3-methyl uracil (*12,13*). Excretion takes place in the kidneys, and from 1 – 6 % of the caffeine ingested is excreted without being metabolized (*11-13*)

Caffeine Entering Boston Harbor. The Massachusetts Water Resources Authority's (MWRA) Deer Island sewage treatment facility handles the wastewater from about 2,000,000 people in the Greater Boston area. Given the estimated daily caffeine consumption and human metabolism efficiency described above, an expected 4.1 – 24.7 kg d^{-1} of caffeine should enter the Deer Island Sewage Treatment Plant. Using an average daily flow of 1.01×10^9 L, 4,100 – 24,700 ng L^{-1} should be present in the sewage entering the plant. A single sample taken in May of 1998 had a caffeine concentration of 15,200 ng L^{-1}, suggesting that indeed domestic wastewater resulting from human consumption is the major source of caffeine. This data is summarized in Table I.

The caffeine concentration of the sewage effluent after 2° treatment being discharged into Boston Harbor was 3,210 ng L^{-1}. This value represents a 78% removal of caffeine from the sewage stream as it passes through the Deer Island facility, which compares favorably to the 80% target set for the removal of total organic material from the effluent (Wenger, E., MWRA, personal communication, 1998).

Table I. Entering Boston Harbor

Estimated population	2,000,000
Daily per capita caffeine consumption	206 mg (*10*)
Caffeine percentage excreted	1 – 6% (*11-13*)
Amount of caffeine excreted daily	4.1 – 24.7 kg
Average daily sewage volume entering plant	1.01×10^9 L[a]
Estimated caffeine concentration in influent	4,100 – 24,700 ng L^{-1}
Measured caffeine concentration in influent	15,200 ng L^{-1}

[a](Rhode, S., MWRA, personal communication, 1998)

Caffeine concentrations measured at six sites in Boston Harbor ranged from 76 - 92 ng L^{-1} (Figure 1). A concentration of 163 ng L^{-1} was measured in the Charles River Basin, and a concentration of 327 ng L^{-1} was measured at the Deer Island Outfall at the mouth of Boston Harbor. Figure 2 is a plot of caffeine concentration versus salinity for four sites in Boston's Inner Harbor, where geographical considerations offer a simpler mixing regime than the whole harbor. These data seem to indicate that caffeine behaves conservatively (with respect to seawater) as highly caffeinated freshwater mixes with "clean" seawater, indicating a possible use of caffeine as a tracer of anthropogenic inputs into marine systems. However, further work is needed to confirm this conclusion.

17α-Ethynilestradiol. Recent attention has focused on natural and synthetic estrogens in sewage effluents emptying into natural waters (*4,14-16*). These compounds include the natural estrogens estrone and 17β-estradiol, as well as 17α-ethinylestradiol, a synthetic estrogen. It has also been suggested that natural and synthetic estrogens are responsible for the estrogenic activity observed in bioassays of sewage effluents, not other suspected endocrine disruptors (*15*). It has also been suggested that bacteria present in sewage treatment facilities may be deconjugating the inactive metabolites of natural and synthetic estrogens present in the effluent into the parent, active compounds, which are then being detected by their estrogenic activity (*15*). The effectiveness of the synthetic estrogen 17α-ethynilestradiol, an oral contraceptive (*4,16*), as an endocrine disruptor has been previously documented (*5*). 17α-ethynilestradiol has been used by itself and in conjunction with diethylstilbestrol to feminize male tilapia in fish farming operations (*17,18*).

17α-Ethynilestradiol was not detected in Deer Island sewage effluent at a calculated detection limit of 74 ng L^{-1}. The detection limit (signal to noise ratio of 3:1) was calculated by multiplying the concentration of the low calibration standard by the extract volume and dividing by the sample volume as in the equation below:

$$2.127 \text{ ng } \mu L^{-1} \times (245 \ \mu L \div 7.08 \text{ L}) = 74 \text{ ng } L^{-1}.$$

This calculated detection limit is considerably higher than the 0.2 ng L^{-1} detection limit obtained by other investigators (*15*). This elevated detection limit is due to the volume of sample extracted and the complex matrix of the effluent extracts. The complex matrix of extractable organics may have masked the presence of 17α-ethynilestradiol

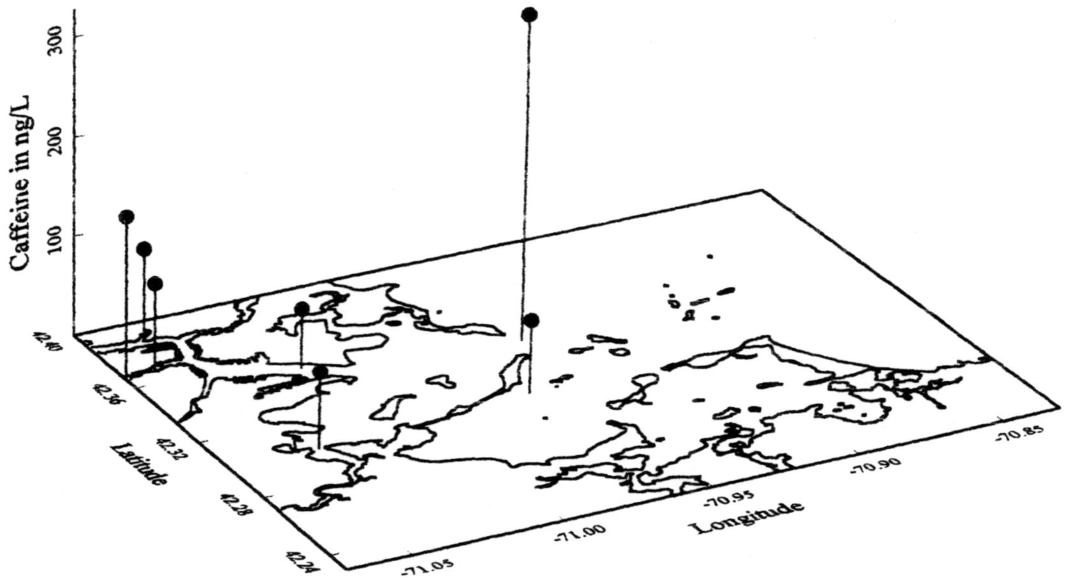

Figure 1. Caffeine Concentrations in Boston Habor

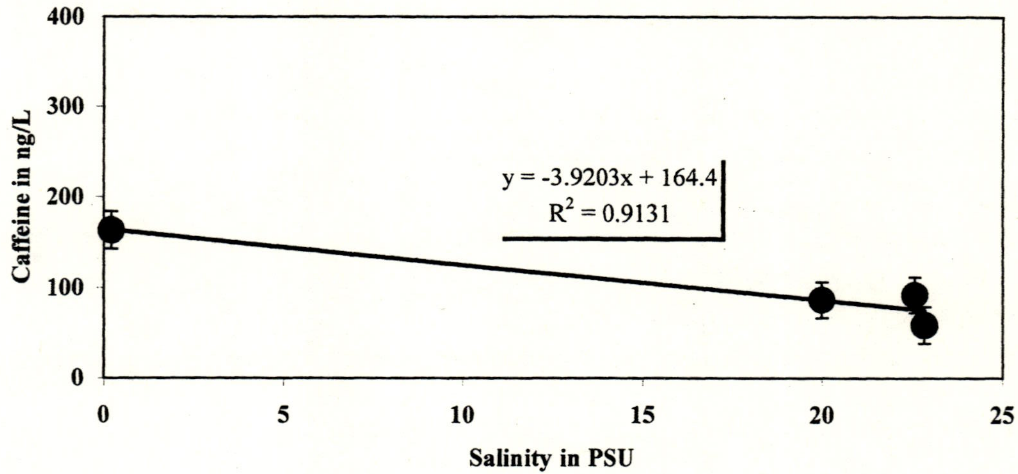

Figure 2. Caffeine Concentration vs. Salinity in Boston's Inner Harbor.

which has been found in sewage effluents at 0.2 – 7.0 ng L^{-1} (*15*). In addition, no sample cleanup (i.e. high pressure liquid chromatography (HPLC)) was performed on the effluent extract in this study, which made concentration of the extract to volumes less than 245 μL impossible and added to the matrix problems noted above.

Although 17α-ethynilestradiol was not detected in effluent discharged in to the Boston Harbor, some worthwhile conclusions can be made. Currently the sewage effluent entering Boston Harbor at the Deer Island Outfall undergoes a 14:1 initial dilution (*19*). Based on a detection limit of 74 ng L^{-1} and the dilution mentioned above the concentration of 17α-ethynilestradiol in Boston Harbor would be 5.3 ng L^{-1}. The new outfall pipe scheduled to open in Massachusetts Bay in 1999, is designed for an initial dilution of 150:1 (*19*), resulting in seawater concentrations of < 0.49 ng L^{-1}. This concentration is at the lower end 0.1 – 10 ng L^{-1} reported to induce estrogenic activity in male trout (*5*).

Other Anthropogenic Compounds. Several other anthropogenic compounds were tentatively identified in sewage effluents and Boston Harbor water samples. Ibuprofen was identified in samples of Deer Island sewage influent and effluent collected in November of 1997. Concentrations for ibuprofen were about 340 ng L^{-1} for the influent and about 4.1 ng L^{-1} for the effluent. Isomers of nonylphenol, a suspected endocrine disruptor, were also detected at significant levels (signal to noise ratio of GC peak > 10:1) in sewage influent, effluent, and Boston Harbor seawater samples from May of 1998. Concentrations of nonylphenol up to 5 μg L^{-1} in Boston Harbor have previously been reported (*20*). Other compounds detected in sewage effluents included benzophenone, a fixative used in soaps and the manufacture of antihistamines and insecticides (*21*), the sunscreen oxybenzone (*21*), and the insect repellent N, N-diethyltoluamide (Deet). Concentrations of these compounds were in the 0- 20 ng L^{-1} range for seawater samples.

Conclusions

Although water quality in Boston Harbor, and natural waters nationwide, has improved over the last 25 years, trace levels of a number of anthropogenic compounds can be detected and measured in marine systems impacted by human activities. While 17α-ethynilestradiol was not detected entering Boston Harbor, a number of anthropogenic compounds were detected at levels from < 5 – 340 ng L^{-1}, ranging from sunscreens to pharmaceuticals to potential endocrine disruptors. Some of these compounds, like caffeine, while not particularly toxic, do have value as potential traces of anthropogenic inputs into marine systems. Others, like the nonylphenols, are suspected endocrine disruptors, and therefore may be having an adverse impact on exposed aquatic organisms. The impact compounds like benzophenone, oxybenzone or deet may be having are unknown at this point. Accurately measuring these compounds in complex matrices such as sewage effluents and seawater presents distinct analytical challenges and represents only the first step to determining the fate and effects of pharmaceuticals in the marine environment.

Acknowledgments

The authors would like to thank Steve Rhode and Ethan Wenger at MWRA for the sewage samples and flow data. The authors would also like to thank Steve Rudnick and Penny Vlahos at UMass Boston, and Rich Restucci at Battelle Ocean Sciences for their assistance with various aspects of this study.

References

1. *U.S. Bureau of the Census, Statistical Abstract of the United States: 1998 (118th Edition)*, Washington, D. C., 1998, pp.1020.
2. Jobling, S.; Nolan, M.; Tyler, C. R.; Brighty, G.; Sumpter, J. P. *Environ. Sci. Technol.* **1998,** *32*, 2498.
3. Hun, T. *Water Environ. Technol.* **1998,** *10*, 17.
4. Routledge, E. J.; Sheahan, D.; Desbrow, D.; Brighty, G. C.; Waldock, M.; Sumpter, J. P. *Environ. Sci. Technol.* **1998,** *32*, 1559.
5. Purdom, C. E.; Hardiman, P. A.; Bye, V. J.; Eno, N. C.; Tyler, C. R.; Sumpter, J. P. *Chem. Ecol.* **1994,** *8*, 275.
6. Barone, J. J.; Roberts, H. R. *Fd. Chem. Toxic.* **1996,** *34*, 119.
7. Sullivan, J. L.; Knodel, L. C. *Toxic Sub. Mech.* **1996,** *15*, 145.
8. Dulloo, A. G., Seydoux, J., Girardier, L. *Am. J. Physiol.* **1994,** *267*, E801
9. Grand, A. N.; Bell, L. N. *J. Am. Diet. Assoc.* **1997**, *97*, 179.
10. Graham, D. M. *Nutrition Reviews.* **1978,** *36*, 97.
11. Chou, T. N.; Benowitz, N. L. *Comp. Biochem. Physiol.* **1994,** *109C*, 173.
12. Pozniak, P. C. *J. Amer. Diet. Assoc.* **1985,** *85*, 1127.
13. Leonard, T. K.; Watson, R. R.; Mohs, M. E. *J. Amer. Diet. Assoc.* **1987**, *87*, 1048
14. Shore, L. S.; Gurevitz, M.; M. Shemesh. *Bull. Environ. Contam. Toxicol.* 1993, *51*, 361.
15. Desbrow, C.; Routledge, E. J.; Brighty, G. C.; Sumpter, J. P.; Waldock, M. *Environ. Sci. Tech.* **1998,** *32*, 1549.
16. Panter, G. H.; Thompson, R. S.; Sumpter, J. P. *Aquat. Toxic.* **1998**, *42*, 243.
17. Melard, C. *Aquaculture* **1995**, *130*, 25.
18. Potts, A. C.; Phelps, R. P. *J. Appl. Ichthyol.* **1995,** *11*, 111.
19. Massachusetts Water Resources Authority, *The State of Boston Harbor 1996: Technical Report No. 97-5.* Boston, MA, 1997, pp. 8.
20. Baldwin, W. S.; Graham, S. E.; Shea, D.; LeBlanc, G. A. *Environ. Toxic. Chem.* **1997**, *16*, 1905.
21. *The Merck Index – 12th Edition*; Budavari, S., Sr. Ed.; Merck Research Laboratories Division: Whitehouse Station, N. J., 1996.

Chapter 9

Identifying Endocrine Disruptors by High-Resolution Mass Spectrometry

Andrew H. Grange[1] and G. Wayne Sovocool

NERL, Environmental Sciences Division, U.S. Environmental Protection Agency, P.O. Box 93478, Las Vegas, NV 89193–3478

Identification of trace amounts of compounds found in the environment, including endocrine disruptors, is difficult. Two analytical tools developed by the U.S. EPA's Environmental Sciences Division, Mass Peak Profiling from Selected Ion Recording Data (MPPSIRD) and a Profile Generation Model (PGM), utilize high resolution mass spectrometry to determine elemental compositions of molecular and fragment ions. These tools provide structural details of molecules and limit the identity of compounds to a manageable number of isomers. Compounds tentatively identified through literature searches can be confirmed by comparison of analyte mass spectra and retention times with those of standards. The enhanced speed and sensitivity provided by MPPSIRD allow elemental compositions to be determined for ions from compounds entering the mass spectrometer as chromatographic peaks in amounts too small for FTIR and NMR to be used. Application of MPPSIRD and the PGM for solving real world environmental contamination problems is reviewed, and a detection limit for ethinylestradiol with probe introduction is determined.

The EPA is currently interested in human and ecosystem exposure to endocrine disruptors (1) – compounds that interfere with endogenous hormone systems. Possible endocrine disruptors in the environment include certain pesticides, industrial by-products, and pharmaceuticals. Such chemicals could be found in surface water or wells as a result of agricultural run off (2), leaching from contaminated sites (3,4), or in treated wastewater discharged from urban areas (5).

Biologically based assays are often used to screen for the presence of endocrine disruptors. While these tests are sensitive, they are non-specific and not inclusive. Cross-reactivity is common but can be an advantage when searching for unidentified endocrine disruptors.

Once a positive result is obtained from a biologically based assay, it is important to identify the compound or compounds responsible for the reactivity. Toxicological studies can then be performed for the identified chemicals. Also, when screening for a specific compound, possible cross-reactivity suggests that a more specific confirmatory technique should be applied to some fraction of the samples showing a positive result for a target analyte.

Limitations of Low Resolution Mass Spectrometry (LRMS)

When possible, compounds to be identified are introduced into the ion source of a mass spectrometer from a gas chromatography (GC) column or from a direct insertion probe and are ionized by electron impact (EI). The electron energy most commonly used, 70 ev, is sufficient to induce fragmentation of the molecular ion. Any remaining molecular ions and the resulting fragment ions provide the mass spectrum of a compound, which is a histogram of ion abundances as a function of mass-to-charge (m/z) ratio. A compound can be identified by matching its mass spectrum with one in a reference library of mass spectra. The low mass resolution provided by quadrupole or ion trap mass spectrometers is not a limitation when only one good library match is found.

If several plausible matches are found, however, low mass resolution is insufficient to distinguish between matches for compounds with different elemental compositions. In addition, such libraries have a limited number of entries and often no matches are found. Missing from mass spectral libraries are most waste products from industrial syntheses. Only half of the 100 drugs most prescribed in the US in 1997 (6) were found in the NIST library (7). To be included in the library, gas chromatographic or direct probe introduction of a drug into the ion source must provide a mass spectrum induced by electron impact ionization. Many drugs and environmental contaminants, however, are too polar, ionic, thermolabile, or involatile to be introduced by GC or probe. Liquid sample introduction using electrospray ionization (ESI) or atmospheric chemical ionization (APCI) are most often used to analyze such compounds (8). These low-energy ionization processes provide little or no fragmentation unless additional voltages are applied to components of the sources. No commercial mass spectral libraries for ESI or APCI are available. Clearly, LRMS is insufficient for identifying many compounds found in the environment.

High Resolution Mass Spectrometry (HRMS)

For double focusing mass spectrometers, mass resolution is defined as $R = \bar{M}/\Delta M$ for a 10% valley between partially overlapping profiles of equal height, where $\bar{M}$ is the average center mass of the two profiles and ΔM is the mass difference between the maxima of the two profiles (9). At mass resolutions of 10,000 or more, the exact mass of the molecular ion formed from a compound can be determined to within a few ppm. The error in the measurement expressed as ppm is $(M_e - M_t)/M_t \times 10^6$, where M_e is the experimentally determined exact mass and M_t is the theoretical exact mass. If the ion's mass is less than 150 amu and contains only C, H, N, O, P, or S atoms, its exact mass usually corresponds to a unique elemental composition. Structural features of the molecule can be deduced from the elemental compositions of the molecular and fragment ions.

Historically, use of HRMS for characterizing or identifying environmental contaminants has been limited. The number of elemental compositions possible for an exact mass increases rapidly with an ion's mass, the number of elements considered, and the error limits of the exact mass determination. Most environmental contaminants have molecular weights greater than 150 amu, and for a given exact mass, multiple compositions are possible.

Using double focusing mass spectrometers, exact masses were determined for analyte ions using peak matching techniques or electric full scans with high mass resolution. Both techniques were too slow to acquire the necessary data for compounds that eluted into a mass spectrometer from a GC over an interval of about 10 sec.

Endocrine disruptors present at very low concentrations in lakes, streams, or drinking water could impact biota or influence human fetal development (10). Thus, low detection limits are desirable. Unfortunately, there is an inverse relationship between mass resolution and sensitivity; as the resolution is increased by narrowing the entrance and exit slits to provide accurate exact masses and to increase selectivity, fewer ions reach the detector.

To overcome the scan speed and sensitivity limitations, the EPA's Environmental Chemistry Branch (ORD/NERL, Las Vegas) developed a HRMS data acquisition technique, Mass Peak Profiling from Selected Ion Recording Data (MPPSIRD) (11,12). For full scans, all m/z ratios across a wide mass range are monitored sequentially, starting at one end of the mass range and finishing at the other. With selected ion recording (SIR), a limited number of m/z ratios are monitored for a longer time, and jumps are made between the m/z ratios of interest. A mass resolution of 10,000 and SIR are used in EPA Methods 8290 (13) and 1613 (14) for quantitation of polychloro-dibenzofurans and polychloro-p-dibenzodioxins to achieve low detection limits while discriminating against interferences. The m/z ratios at the maxima in the mass peak profiles of analyte and calibrant ions are monitored. Two profiles are monitored for each analyte, and the ratio of the signals indicates a lack of major interferences when it falls within 15% of the expected value. Beyond these methods, researchers use abbreviated full scans across narrow mass ranges that include the full mass peak profile of each analyte ion to reveal interferences (15,16). An incorrect exact mass or deviation of the profile from a Gaussian shape results from interferences. Increased sensitivity is realized when multiple m/z ratios are used with SIR to plot the profiles of analytes for the same purpose (17). These data acquisition techniques provided a quality assurance tool for dioxin analyses. At the ESD, automated procedures were developed to prepare SIR descriptors (which contained a list of the m/z ratios to be monitored), acquire SIR data, plot profiles, and calculate exact masses and relative abundances. From these data, elemental compositions of ions were determined.

Compared with full scanning, MPPSIRD provided a 170-fold lower detection limit (6 fg) for a tetrachlorobiphenyl eluting from a GC with 10,000 resolution (11). At 20,000 resolution, MPPSIRD provided a six-fold faster cycle time (0.8-1.0 sec/cycle) compared with conventional mass spectrometric scans. Consequently, data can be acquired for ions from compounds that enter the ion source as chromatographic peaks. In addition, routine use of up to 20,000 resolution provides excellent discrimination against interferences having the same nominal (integer) masses, but different elemental compositions than the ion under study (18).

To increase the largest m/z ratio for which unique elemental compositions could be determined from 150 to 600 amu, a Profile Generation Model (PGM) (19) was written in QuickBasic to consider mass peak profiles with m/z ratios 1 and 2 amu larger than that of the ion of interest. Associated with the molecular ion (M), are M+1 and M+2 profiles that arise from the presence of heavier isotopes. If M contains C, H, N, O, P, or S, then ^{13}C, ^{2}H, ^{15}N, ^{17}O, ^{18}O, ^{33}S, or ^{34}S are considered. Ions containing one atom of an isotope heavier by 1 amu than the most common isotope contribute to the M+1 profile, while ions having one atom of an isotope heavier by 2 amu or two atoms of isotopes heavier by 1 amu contribute to the M+2 profile. In addition to the exact mass of the M profile, the model calculates the exact masses of the M+1 and M+2 profiles and the abundances of the M+1 and M+2 profiles relative to the M profile. Comparison of calculated and experimental values provides four additional criteria for distinguishing between possible elemental compositions of an ion.

MPPSIRD was developed using a VG 70-250SE double focusing mass spectrometer. Recently, a new Finnigan MAT 900-trap hybrid mass spectrometer was installed in our laboratory. The VG manual uses the term "selected ion recording" to describe acquiring data only at discrete m/z ratios, while Finnigan MAT's term for the same technique is "multiple ion detection" (MID). To maintain continuity in the published literature, we continue to call the technique MPPSIRD even when data are acquired with the Finnigan instrument.

MPPSIRD

In Figure 1a are shown ion chromatograms corresponding to the maxima in the molecular (M), M+1, and M+2 mass peak profiles for an analyte and for the profiles of two calibration ions from the calibrant, perfluorokerosene (PFK). The areas under the chromatographic peaks and those for m/z ratios acquired at 10-ppm intervals on either side of the maxima were plotted to provide the partial profiles in Figure 1b. The analyte was volatilized from a heated probe into the ion source, while a constant level of PFK entered the ion source from a heated reservoir. The baseline excursions in the ion chromatogram for the calibrant occurred when the shield voltage in the ion source was set to 0 V for 5 sec. Automated procedures determined the area between the baseline excursions, which defined a simulated chromatographic peak for the ever-present calibration ion. Integration of the areas under the wide peaks in Figure 1a obtained as the analyte was volatilized from a heated probe improved the signal-to-noise ratio and contributed to the enhanced sensitivity of MPPSIRD. Before data were acquired for each cycle, the data system software locked onto the center mass of the profile of the first calibration ion to compensate for calibration drift, thereby avoiding loss of resolution in the plotted mass peak profiles despite integration of the peak areas over a long interval.

The five partial profiles in Figure 1b were acquired with a mass resolution of 10,000. The two smaller displays show partial profiles for the calibration ions, which were each plotted from five m/z ratios. The three partial profiles for the analyte were each plotted from seven m/z ratios that span 70% of the mass range of each profile. Monitoring partial rather than full profiles permits study of five ions simultaneously using all 31 m/z ratios available in a MID descriptor. Exact masses of the M+1 and M+2 profiles and their abundances relative to the M profile were determined from the partial profiles. The weighted average of the areas defining each partial profile provided its exact mass, and ratios of the sums of the areas yielded the relative abundances.

Files to be executed by the MAT 900 data system were prepared by a Lotus 123 (v2.2) program on the personal computer (PC) that contains the ion trap data system after the user specified the center masses of the analyte profiles, the mass resolution used to set the mass increments, and the data file name. These files were transferred to the mini-computer that controls the Finnigan MAT 900 using Ftp, a DOS-based file transfer program supplied by Finnigan, where their execution prepared the MID descriptors, induced the baseline excursions, and, after data acquisition, prepared an ASCII file with the m/z ratios and chromatographic peak areas. After transfer of the ASCII file back to the PC, Lotus procedures prepared and plotted the profiles using WordPerfect 5.1. DOS batch files provided automated transition between Lotus, Ftp, and WordPerfect on the PC.

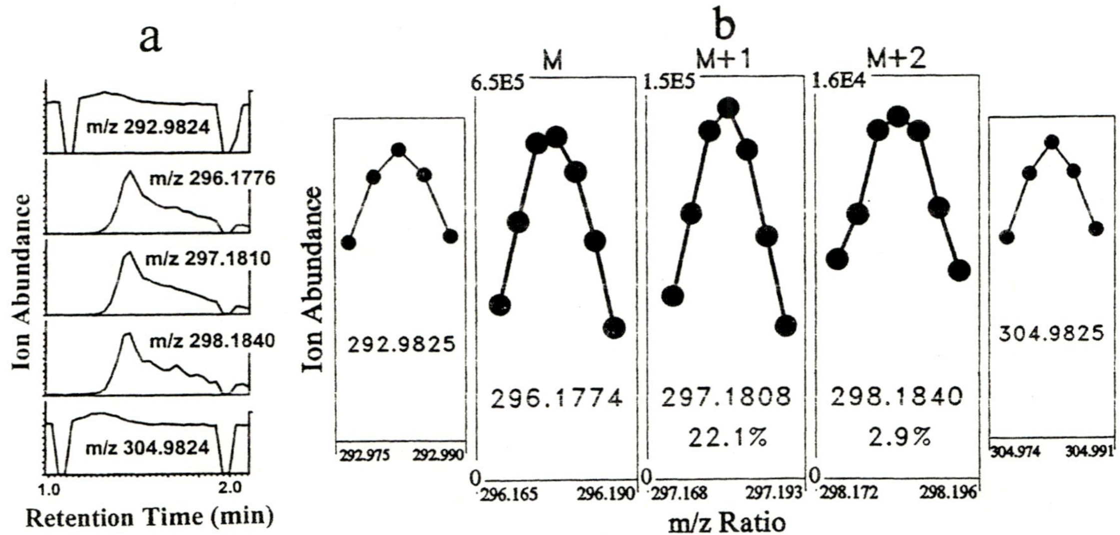

Figure 1. Ion chromatograms (a) from which the areas under the chromatographic peaks correspond to the maxima in the five partial mass peak profiles in (b).

PGM

For an unidentified ion, an exact mass is determined for the full M profile at 10,000 resolution. This exact mass and the possible elements comprising the ion are entered into the PGM, and a table listing all possible compositions of the ion within the error limits of the mass determination is provided. (To shorten the list, at least 1/3 of the mass is usually assumed to be from C atoms). Based on any available knowledge of the sample, the user selects from the list a hypothetical composition and enters the corresponding exact masses of the M, M+1, and M+2 profiles into the automated procedure that prepares the MID descriptor to acquire the data pictured in Figure 1.

The five experimental values are then entered into the PGM to provide a table in which ranges of acceptable values that account for errors associated with the use of partial profiles (19) are calculated and compared to the observed values. The mass defects (fractional masses) are listed rather than the full 7-digit exact masses to allow a larger font. The last line in Table 1 lists average experimental values from triplicate determinations. An "X" next to an entry indicates disagreement with the predicted value, and the composition is rejected. In this example, as is generally the case, only one composition, $C_{20}H_{24}O_2$, was consistent with the five measured values. Automatic interpretation by the PGM of the data acquired using MPPSIRD provides compositions of ions with m/z ratios up to 600 amu. This capability expands the utility of HRMS to a much larger fraction of environmentally significant compounds.

Table 1. *Comparison of Experimental and Theoretical Values**

m/z 296.1773 ± 6 ppm — Resolution: 10000

Composition	M	M+1	M+2	%M+1	%M+1 Range	%M+2	%M+2 Range
$C_8H_{20}ON_9F_2$	.1759	.1777 X	.1796 X	11.9	(9.0-13.9) X	0.9	(0.6-1.0) X
$C_8H_{23}N_9SF$	.1781	.1798	.1748 X	12.7	(10.0-15.4) X	5.0	(0.9-5.2)
$C_8H_{26}N_8P_2$	.1756	.1776 X	.1794 X	11.6	(8.8-13.5) X	0.6	(0.4-0.8) X
$C_9H_{24}ON_7PF$	.1764	.1787 X	.1808 X	12.4	(10.0-14.6) X	0.9	(0.6-1.1) X
$C_9H_{27}N_7PS$	.1786	.1807	.1755 X	13.2	(10.9-16.0) X	5.0	(1.1-5.5)
$C_{10}H_{22}O_2N_6F_2$	.1772	.1797	.1819 X	13.3	(11.1-15.4) X	1.2	(0.9-1.5) X
$C_{10}H_{26}O_3N_5S$	.1756	.1781 X	.1734 X	13.8	(10.9-16.0) X	5.4	(1.0-5.0)
$C_{10}H_{28}ON_5P_2$	.1769	.1796	.1819 X	13.0	(10.5-15.1) X	1.0	(0.7-1.2) X
$C_{11}H_{26}O_2N_4PF$	.1777	.1806	.1829	13.8	(11.6-16.1) X	1.3	(1.0-1.6) X
$C_{12}H_{24}O_3N_3F_2$	.1786	.1816	.1839	14.7	(12.4-17.2) X	1.6	(1.3-2.0) X
$C_{12}H_{28}O_4N_2S$	.1770	.1800	.1753 X	15.2	(12.6-17.7) X	5.6	(1.6-6.0)
$C_{12}H_{30}O_2N_2P_2$	.1783	.1814	.1840	14.4	(12.4-16.5) X	1.4	(1.1-1.6) X
$C_{13}H_{21}ON_6F$	.1761	.1788 X	.1812 X	16.5	(13.8-19.1) X	1.5	(1.1-1.8) X
$C_{13}H_{24}N_6S$	.1783	.1809	.1760 X	17.3	(14.5-20.3) X	5.3	(1.5-6.0)
$C_{13}H_{29}N_2S_2F$	.1756	.1785 X	.1724 X	16.9	(13.9-19.5) X	9.6	(0.9-7.9)
$C_{14}H_{25}ON_4P$	.1766	.1796	.1822 X	17.1	(14.3-19.7) X	1.6	(1.3-1.9) X
$C_{15}H_{23}O_2N_3F$	.1774	.1805	.1832	17.9	(15.0-20.9) X	1.9	(1.5-2.3) X
$C_{15}H_{24}OF_4$	.1763	.1797	.1828	17.1	(14.7-19.5) X	1.6	(1.3-1.8) X
$C_{15}H_{27}SF_3$	.1786	.1818	.1762 X	17.8	(15.0-20.9) X	5.2	(1.7-6.0)
$C_{16}H_{27}O_2NP$	.1779	.1813	.1841	18.5	(15.6-21.6) X	2.0	(1.6-2.4) X
$C_{17}H_{25}O_3F$	.1788	.1822	.1850	19.4	(16.7-22.1) X	2.4	(1.9-2.8) X
$C_{18}H_{22}ON_3$	.1763	.1794	.1823	21.2	(17.7-24.4)	2.3	(1.9-2.8) X
$C_{20}H_{24}O_2$	.1776	.1810	.1841	22.7	(19.5-25.9)	2.8	(2.4-3.3)
Observed:	.1773	.1806	.1841		22.3		3.1

*Possible compositions based on the exact mass of the molecular ion and the error in its determination; calculated mass defects for the M, M+1, and M+2 partial profiles; and calculated relative abundances for the M+1 and M+2 partial profiles assuming the profiles' center masses for each composition are used in the MID descriptor, or in parentheses, the range expected if the center masses of the hypothetical (last) composition are used.

Municipal Well Pollutants

In a recent example, MPPSIRD and the PGM were used to determine the elemental composition of several isomers found in a municipal well that serviced Toms River, NJ, where an increased incidence of childhood cancer had been observed (3). No mass spectra similar to the one in Figure 2a were found in the NIST library. To illustrate the reduction in possible compounds that results from determining the elemental composition of the molecular ion, the library was searched for m/z 210 and for its composition, $C_{14}H_{14}N_2$. For this mass, 342 entries were found, but for the composition, the possibilities were reduced to 13. Determination of compositions of fragment ions and neutral losses based on their exact masses and the elemental limits established by the molecular ion revealed whether fragment ions contained 0, 1, or 2 N atoms. This information and review of library mass spectra for smaller molecules with hypothesized sub-structures suggested that the isomers contained a tetralin core with a cyano group and an α-cyanoethyl or β-cyanoethyl group attached to it. A literature search for the limited number of possible isomers that remained revealed that polymer synthesis of styrene and acrylonitrile probably produced the isomers. Three of the five isomers in the well water provided the same mass spectra and GC retention times as three isomers in a sample from a similar industrial synthesis. The concentration of the isomers in the well water extract was too low for study by FTIR or NMR.

This example illustrates that information in addition to fragment and molecular ion compositions is generally required to identify compounds. However, the elemental compositions greatly limit the number of isomers that are possible and the time required to search the literature for such isomers.

Characterization of a Superfund site sample

A small amount of a tar-like sample from a Superfund site in W. Virginia was dissolved in methylene chloride and injections were made onto a 30-m, 0.25-mm i.d., 0.25-μm film thickness, RT_x-5 column (18). The GC temperature program started at 40°C for 3 min and increased by 8°C/min to 320°C, where it remained for 15 min. Compounds were investigated that were responsible for 47 peaks in the total ion chromatogram (TIC), 35 of which are shown in Figure 3. A few low-mass compounds were identified from library matches of mass spectra alone, but these compounds accounted for only a small fraction of the total signal from the sample. Using a mass resolution of 20,000, MPPSIRD and the PGM were used to determine the elemental compositions of the presumed molecular ion in each mass spectrum for the remaining chromatographic peaks. The composition labels in Figure 3 demonstrate that most of the ions contained at least one N atom and one S atom. In addition, the ion chromatogram in Figure 3 (lower trace) was obtained using 20,000 resolution for the $C_8H_7NS^+$ ion (149.0299 amu) characteristic of alkylbenzothiazoles. This mass resolution discriminated against the common $C_8H_5O_3^+$ ion (m/z 149.0239) from phthalates, although no phthalates were present. The same retention time for peaks in the TIC and the ion chromatogram indicated that many of the compounds contained the benzothiazole moiety. Benzothiazoles are used by the dye and rubber industries. These results were consistent with the known source of this contamination. This study provided elemental compositions for significant ions formed from compounds that accounted for most of the sample's mass and thereby adequately characterized the sample for determining the origin of the waste.

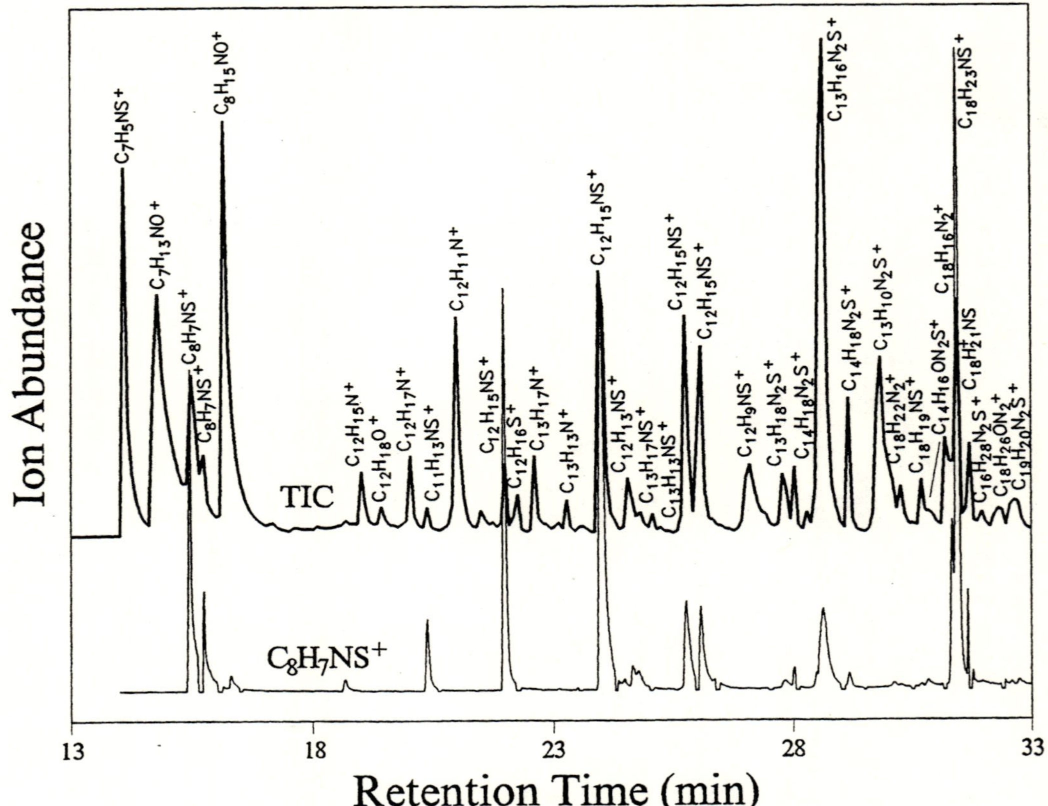

Figure 2. Mass spectra for an isomer in a well water extract (a) and for a standard with the same retention time (b).

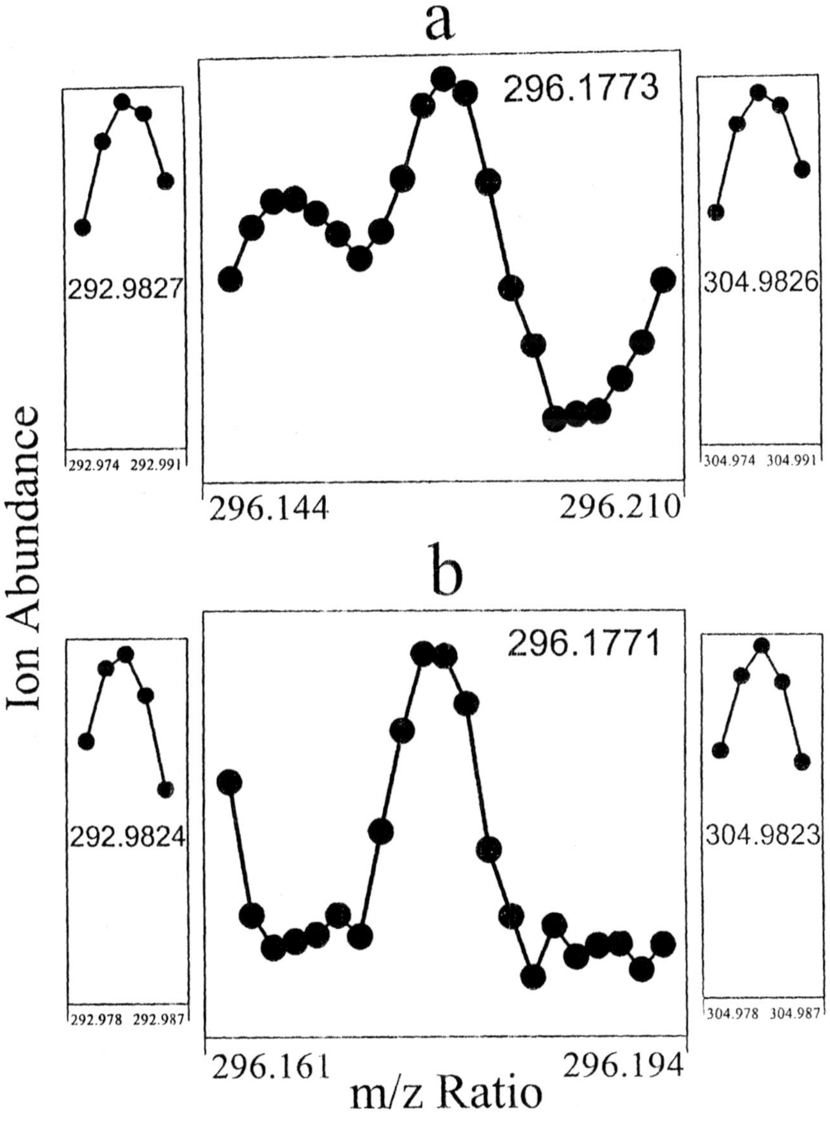

Figure 3. Portion of a total ion chromatogram (upper trace) acquired with 1000 resolution and an ion chromatogram for the $C_8H_7NS^+$ ion (m/z 149.0299) from alkylated benzothiazoles (lower trace) acquired with 20,000 resolution.

This sample demonstrated the importance of the wide linear dynamic range afforded by double focusing mass spectrometers. The areas under the peaks in the TIC in Figure 3 differed by more than a factor of 10, and the area under the M+2 profile was as small as 1% of the area under the M profile for ions not containing S atoms.

Probe Introduction

For the same sample (18), elemental compositions were determined for several higher-mass compounds that did not elute from the GC column by introducing the sample extract into the ion source with a heated direct insertion probe. High mass resolution was used to separate ions in the mass domain, where separation of neutral compounds prior to ionization by GC had failed. This work suggested that using a heated probe for rapid screening of target analytes, including endocrine disruptors such as ethinylestradiol (EED), could be feasible. EED is a potent endocrine disruptor, routinely used for birth control. Treated wastewater entering Lake Mead, the drinking water supply for Las Vegas, could possibly contain EED (20).

In Figure 1b are shown partial profiles obtained for 10 ng of EED volatilized from the probe. The average values from three such determinations were entered into the PGM to provide Table 1. The elements considered were C, H, N, O, P, S, and F. If such results were obtained for a sample, investigation of characteristic fragment ions would provide additional evidence for the presence of EED. Because data from both the M+1 and M+2 profiles are needed to reject all but the correct composition, the detection level is about 50 times higher than one based on the molecular ion alone. However, the correct exact mass for $C_{20}H_{24}O_2$ would still provide a more specific screen for EED than a biologically based assay and would be susceptible to fewer, and different, interferences.

To determine a detection limit based on the molecular ion alone, the detector sensitivity was increased and 1, 10, and 100 pg amounts of the analyte were inserted on the probe. MPPSIRD was used to plot the full M profile. As the concentration of analyte introduced into the source was reduced, the importance of interferences with the same nominal mass increased. For the full profile in Figure 4a acquired with 10,000 resolution, the profile of a lower-mass interference was observed to partially overlap that of the analyte. For lower concentrations, the signal due to the analyte would be obscured by the interference, and an accurate mass for the analyte could no longer be obtained. For the profile in Figure 4b acquired with 20,000 resolution, this interference was no longer important. However, a 3-fold loss in ion abundance accompanied the 2-fold increase in resolution and increased the detection limit. Although 10 pg was readily detected with 20,000 resolution, a laboratory contamination level of the same magnitude was discovered. A molecular or fragment ion with the same elemental composition as the analyte ion was produced from the contaminant. EED had been weighed out a few feet from the instrument a few weeks earlier and was probably responsible for the contamination. Even so, detection of 10 pg above the contamination level corresponds to a detection limit of 1 ppt if the EED in 1 L of water were concentrated into a 0.1 mL extract followed by injection of 1 μL of the extract into the probe tip capillary for analysis.

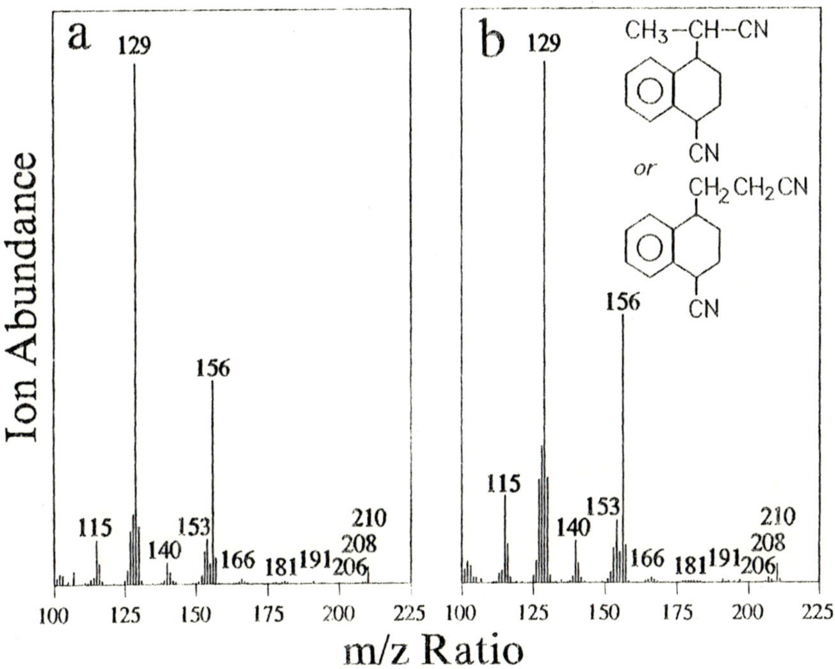

Figure 4. Full mass peak profiles obtained with (a) 10,000 resolution and (b) 20,000 resolution for ethinylestradiol.

Current Work

These studies suggest that MPPSIRD and the PGM will be powerful tools for identifying endocrine disruptors. Initial attempts to identify such compounds will focus on analysis of tertiary treated wastewater that flows into Lake Mead, the drinking supply for Las Vegas. Concentrations of consumer product wastes, pharmaceuticals, lawn and garden pesticides, and their metabolic or waste treatment products should be much greater before dilution in the lake. In another chapter of this book, Snyder et al. report separating numerous unidentified compounds in the treated water by HPLC. As the examples above illustrated, using MPPSIRD and the PGM with separation by GC to determine elemental compositions of ions is routine in our laboratory. Before trying to identify the compounds observed by Snyder, we must first interface HPLC to our Finnigan MAT 900S-trap mass spectrometer through the ESI and APCI sources. Exact masses are determined by calibrating between two known masses from calibrants that bracket the mass of the analyte ion. A problem with liquid sample introduction is finding appropriate concentrations of mass calibrants that provide ions under the same conditions as the analytes. No nearly universal calibrant such as PFK, which provides ions every 12-14 amu when gas phase sample introduction is used, is available for liquid sample introduction. To avoid this problem, we have developed a methodology to determine elemental compositions of ions without using mass calibrants. Elemental compositions are determined from relative abundances of M+1, M+2, F+1, and F+2 ions, exact mass *differences* between ions, and the appearance of the M+1 or F+1 profiles (21). This methodology requires infusion of previously separated analytes. We anticipate that using parts of this new approach and a single mass calibrant to achieve lock-on by the data system will allow determination of elemental compositions of ions from compounds entering the ESI or APCI source as HPLC peaks. At that time, we will be prepared to investigate compounds separable by either GC or HPLC after extraction from waste streams, lakes, soils, and other sample matrices.

References

1. Government Performance and Results Act (U.S. EPA), Goal 8, Objective 3, Subobjective 1, Endocrine Disruptors, #8.
2. Halling-Sørensen, B.; Nielsen, S.N.; Lanzky, P.F.; Ingerslev, F.; Lützhøft, H.C.H.; Jørgensen, S.E.; *Chemosphere* **1998**, *36*, 357-393.
3. Grange, A.H.; Sovocool, G.W.; Donnelly, J.R.; Genicola, F.A.; Gurka, D.F.; *Rapid Commun. Mass Spectrom.* **1998**, *12*, 1161-1169.
4. Ferguson, P.L.; Grange, A.H.; Brumley, W.C.; Donnelly, J.R.; Farley, J.W.; *Electrophoresis*, **1998**, *19*, 2252-2256.
5. Buser, H.R.; Muller, M.D.; Theobald, N; *Environ. Sci. & Technol.*, **1998**, *32*, 188-192.
6. http://www.rxlist.com/top200.htm
7. NIST Standard Reference Database 1A, NIST Mass Spectral Search Program and NIST/EPA/NIH Mass Spectral Library, Ver. 1.0, For Use with Microsoft Windows, U.S. Dept. of Commerce, Technology Administration, National Institute of Standards and Technology, Standards Reference Data Program, The NIST Mass Spectrometry Data Center, Gaithersburg, MD 20899, January 1995.
8. Cole, R.B., Ed.; *Electrospray Ionization Mass Spectrometry*, Wiley: New York, 1997, *9*, 324.
9. White, E.A.; Wood, G.M.; *Mass Spectrometry: Applications in Science and Engineering*, Wiley: New York, 1986, 163-164.
10. Colborn, T.; Dumanoski, D.; Myers, J.P.; *Our Stolen Future*, Penguin: New York, 1996.

11. Grange, A.H.; Donnelly, J.R.; Brumley, W.C.; Billets, S.; Sovocool, G.W.; *Anal. Chem.* **1994**, *66*, 4416-4421.
12. Grange, A.H.; Donnelly, J.R.; Sovocool, G.W.; Brumley, W.C.; *Anal. Chem.* **1996**, *68*, 553-560.
13. Method 8290, *Test Methods for Evaluating Solid Wastes*, 1B#2, SW-846, US EPA, Office of Solid Waste and Emergency Response, Washington, CD, 1986.
14. Method 1613, National Center for Environmental Publications and Information, Cincinnati, OH, 1994.
15. Tong, H.Y.; Giblin, D.E.; Lapp, R.L.; Monson, S.J.; Gross, M.L. *Anal. Chem.* **1991**, *63*,1772-1780.
16. Tondeur, Y.; Hass, J.R.; Harvan, D.J.; Albro, P.W. *Anal. Chem.* **1984**, *56*, 373-376.
17. Allan, A.R.; Roboz, J. *Rapid Commun. Mass Spectrom.* **1988**, *2*, 246-249.
18. Grange, A.H.; Brumley, W.C.; *LC•GC* **1996**, *14*, 978
19. Grange, A.H.; Brumley, W.C.; *J. Amer. Soc. Mass Spectrom.* **1997**, *8*, 170-182.
20. Raloff, J.; *Sci. News* **1998**, *153*, 187.
21. Grange, A.H.; Sovocool, G.W.; *Rapid Commun. Mass Spectrom.* **1999**, *13*, 673-686.

Chapter 10

Levels of Selected Nonpersistent Endocrine Disruptors in Humans

Larry L. Needham, Benjamin Blount, Helen Schurz Rogers, John W. Brock, and Dana Barr

Division of Laboratory Sciences, Toxicology Branch, Centers for Disease Control and Prevention, National Center for Environmental Health, 4770 Buford Highway, NE, Mailstop F–17, Atlanta, GA 30341–3724

Introduction

Human endocrine glands secrete a wide variety of hormones. For example, the sex hormones, androgen in males and estrogen in females, are secreted primarily from the testis and ovary, respectively. Other glands that secrete hormones include the adrenal (the adrenal cortical hormone), the thyroid (thyroid hormone), the pituitary (growth hormone), and the pancreas (insulin). They travel in the blood stream in either the free or bound state at concentrations in the parts-per-billion to parts-per-trillion levels to receptors, which are located in cells of internal organs. There, they bind to their receptor and become activated which allows them to transmit signals to DNA to produce functional proteins or to control cell division. Certain environmental chemicals have been reported to alter one or more of these steps, thus affecting normal hormonal function and potentially resulting in adverse health effects, such as cancer, reproductive effects, and behavioral effects, in the exposed animal or its offspring. These environmental chemicals act as hormones and essentially add to a hormonal effect (agonist) or subtract (block) from the effect (antagonist). Because of the potential for adverse health effects, these chemicals have come under recent global scrutiny from the scientific community, politicians, and the public.

We, in the Division of Laboratory Sciences in the National Center for Environmental Health of the Centers for Disease Control and Prevention (CDC), are committed to assess human exposure to a wide variety of environmental toxicants, including many potential endocrine disruptors. These exposure assessments are based on the measurement of the human levels of the environmental toxicant, its primary metabolite, or a reaction product, such as an adduct to DNA (1). The reasons for this approach are discussed later in this chapter. These exposure assessments are used for many purposes including biomonitoring programs, exposure assessments, epidemiological studies, and risk assessments. Each of these four programs may have different objectives. For example, biomonitoring programs of the general population for environmental toxicants may have one or more of the following objectives:

- Establishment of reference range or background levels
- Determination of subpopulations potentially at risk because of high exposure
- Monitoring of trends or changes in human levels, both spatially and temporally
- Identification of new or unrecognized toxicants
- Relationship between reference ranges in humans and wildlife
- Integration of dose assessment with total exposure
- Storage of samples for future research

Various lists of so-called endocrine-disrupting chemicals have been developed. These chemicals have been reportedly shown to cause endocrine disrupting activity in animals (in a laboratory setting or in the wild) and/or in humans. That is, these chemicals have been found under certain exposure conditions to mimic endogenous hormones, and thereby produce similar responses. Again, the lists of these synthetic environmentally-found endocrine disrupting chemicals vary, but they consist of pesticides, including insecticides, herbicides, nematocides, and fungicides, and a growing number of industrial chemicals. Our laboratory has developed the analytical methods to measure many of these endocrine disrupting chemicals in human blood or urine specimens. One such list and the chemicals we have measured in biological specimens are given in Table 1 (2).

These measurements have been performed in support of epidemiological studies and human exposure assessment surveys (3). For the most part, the epidemiological studies relating exposure to adverse health effects have centered on the persistent chemicals, such as the organochlorine insecticides, heavy metals, and polychlorinated dibenzo-p-dioxins, furans, and biphenyls. These chemicals are called "persistent" because of their long half-lives in the environment and in animals, including humans. Thus, exposure of human populations to these chemicals can be assessed for many years after exposure has ceased. This assessment is done by first measuring the levels of these chemicals in the appropriate biological specimen, which is dependent on the environmental toxicant, and then by using pharmacokinetic data, extrapolating back to what levels may have been at a certain point in time. This method was used for assessing occupational exposure to 2,3,7,8-tetrachlorodibenzo-p-dioxin many years after exposure ceased (4). These exposure measurements were then used to relate dioxin levels with cancer.

This approach is consistent in our belief that the best way to assess human exposure to a chemical is to measure the internal dose; that is, the concentration of that chemical, its primary metabolite, or reaction product in a human specimen. However, before we enter into such studies it is advantageous to know if a given chemical is normally found in the general population, and if so, at what levels. Thus we have to determine the so-called "reference range" for a chemical in the general population (5). Such ranges give us information as to the prevalence of exposure of selected chemicals and the normal concentration range found in humans. These ranges also serve as a basis for trend studies, which are designed to determine whether human exposure to a given chemical is increasing or decreasing over a given time period. For toxicological purposes these human reference ranges can help prioritize chemicals for testing based on their prevalence and levels in humans. We are currently providing such human data to the National Toxicology Program (NTP) at the National Institute of Environmental Health Sciences (NIEHS) to help prioritize chemicals for testing in long-term bioassays; many of these chemicals are on the Table 1 list of endocrine disruptors. This prioritization is extremely important because the NTP, which is the nation's largest toxicology program, can initiate only approximately 10 long-term cancer studies and 10 reproductive studies per year. Chemicals found in a high proportion of the population or in high concentrations in certain segments of the population are given a high priority for these toxicological studies (6,7). Specimens for the establishment of these ranges come from control or supposedly "non-exposed" populations in various epidemiologic studies and from national surveys such as the National Health and Nutrition Examination Survey (NHANES)(8) and the pilot phase of the National Human Exposure Assessment Survey (NHEXAS) (9). As a means of introduction, NHANES is a program run by the CDC's National Center for Health Statistics. Our laboratory within NCEH serves as its central laboratory. Biological samples for NHANES III were collected from 30,000 participants from 1988-1994. Environmental toxicants (or metabolites) were measured in various population subsets of NHANES III. These toxicants include selected heavy metals, cotinine (primary metabolite of nicotine), nonpersistent pesticides, and volatile organic compounds (VOCs). To date, we have supplied the NTP with reference ranges to chemicals listed in Table 1. These chemicals include the traditional persistent organochlorine insecticides, polychlorinated biphenyls, dioxins, and furans and heavy metals. In addition, this list includes the nonpersistent pesticides, 2,4-dichlorophenoxyacetic acid, atrazine, pentachlorophenol, carbaryl, and parathion. Two groups of industrial chemicals for which we are developing reference ranges are the phthalates and phytoestrogens. In this forum, we will describe some of our method development and reference range data for selected nonpersistent pesticides, phthalates and phytoestrogens.

Phthalates

Phthalates are a group of chemicals that are diester derivatives of phthalic acid and are used as plasticizers to make certain plastics such as polyvinyl chloride (PVC) more flexible. PVC may contain up to 40% phthalates by weight, but it should be noted that the phthalates are not covalently bound to other ingredients in the PVC. This nonbonding allows the phthalates to be more easily leached from the plastic. Plastics containing phthalates are widely used in consumer products such as imitation leather, rainwear, footwear, upholstery, flooring, tablecloths, shower curtains, food packaging materials, children's toys including teething rings, and in tubing and containers for blood transfusions and kidney dialysis. The most widely used phthalate has been di(2-ethylhexyl)phthalate (DEHP); about 300 million pounds per year were used in the late 1980s/early 1990s but there is some evidence that this amount may be decreasing

Table I. List of Potential Synthetic Endocrine Disruptors

Insecticides (CAS #)	Herbicides (CAS #)	Fungicides (CAS #)	Nematocides (CAS #)	Industrial Chemicals (CAS #)
[a]DDT (50-29-3) /metabolites (72-55-9, 72-5-8)	[a]2,4-D (94-75-7)	[a]Hexachlorobenzene (118-74-1)	Aldicarb (116-06-3)	[a]Cadmium (7440-43-9)
[a]Dieldrin (60-57-1)	[a]2,4,5-T (93-76-5)	Benomyl (17804-35-2)	Dibromochloropropane (96-12-8)	[a]Lead (7439-92-1)
[a]Mirex (2385-85-5)	[a]Alachlor (15972-60-8)	Mancozeb (8018-01-7)		[a]Mercury (7439-97-6)
[a]Chlordane (57-74-9)	[a]Atrazine (1912-24-9)	Maneb (12427-38-2)		[a]PCBs (1336-36-3)
[a]Oxychlordane (27304-13-8)	Amitrole (61-82-5)	Metiram (9006-42-2)		[a]PBBs
[a]Heptachlor (76-44-8) Heptachlor Epoxide (1024-57-3)	Metribuzin (21087-64-9)	Tributyltin oxide (56-35-9)		[a]Dioxins
[a]Trans-nonachlor (39765-80-5)	Nitrofen (1836-75-5)	Zineb (12122-67-7)		[a]Pentachlorophenol (87-86-5)
[a]β-HCCH (319-85-7)	Trifluralin (1582-09-8)	Ziram (137-30-4)		Alkylated Phenols
[a]Toxaphene (8001-35-2)				[a]Phthalates
[a]Lindane (58-89-9)				Styrenes (100-42-5)
Dicofol (115-32-2)				Bisphenol-A (80-05-7)
Methoxychlor (72-43-5)				

Endosulfan (115-29-7)

Pyrethroids (NA)

[a]Carbaryl (63-25-2)

Methomyl (16752-77-5)

[a]Ethyl Parathion (56-38-2)

[a]Reference ranges on these chemicals supplied to NIEHS

although the use of other phthalates may be increasing. The general population is potentially continually exposed to phthalates via the food chain. Humans may be exposed to higher levels of phthalates in certain micro environments described above. Most of the toxicological information available on the phthalates is on DEHP. DEHP has been shown to cause liver tumors in both rats and mice receiving DEHP in their diet throughout their entire life span. The International Agency for Research on Cancer has classified DEHP as a possible human carcinogen. DEHP is not genotoxic (does not bind to DNA) and its carcinogenic mechanism is not well understood; however, many scientists support the hypothesis that DEHP acts as a peroxisome proliferator. DEHP has also been shown to alter the activity of several hepatic enzymes resulting in, for example, a reduced rate of metabolism of ethanol and several therapeutic drugs. Oral administration of DEHP has been shown to produce adverse effects on the male reproductive capacity in laboratory animals; these effects are characterized by decreased organ weights and histological changes in the seminiferous tubules.

In addition to DEHP, toxicological data are emerging for di-n-butyl phthalate (DBP), which is used as a solvent and also as a plasticizer. It has been known for sometime that DBP produces adverse reproductive effects in male rats at high doses. However, more recent data (10) from the NTP showed that while rats orally fed DBP at levels up to 650 mg/kg/day continued to be fertile, their offspring, while receiving the same diet, had extremely poor fertility. This emphasizes the importance of the life stage during exposure - be it gestation, lactation, or puberty. It was of interest to note that examination of the male offspring revealed a number of male reproductive tract malformations, but the female offspring had normal reproductive tract formations. The pattern of response indicated that DBP was not acting as an estrogen but perhaps as a weak antiandrogen although neither DBP nor its primary metabolite binds directly to the androgen receptor. A possible explanation is that DBP may indirectly interfere with androgen signaling pathways during sexual differentiation, possibly by acting directly on fetal testis (Chemical Institute of Industrial Toxicology (CIIT) Activities 18, Sept 1998).

Thus, because of the potential for wide exposure to phthalates and possible human adverse health effects resulting from these exposures and the ability and capability of our laboratory, we are developing the necessary methods for assessing human exposure to these compounds. Our first approach was to measure the parent phthalates in blood; however, with each attempt we faced the difficulty of contamination - this was due to the ubiquity of the phthalates in the environment which resulted in potential contamination of the blood specimens in every step from collection to analysis. Thus, we turned our attention to measuring the urinary metabolites, but this called for some knowledge of the pharmacokinetics of the phthalates. The first step in their metabolism is deesterification of one of the ester groups, resulting in the monoester and the corresponding alcohol. Our methods are based on measuring eight monoester metabolites from eight phthalates - diethyl, dibutyl, di-2-ethylhexyl, dicyclohexyl, dioctyl, diisononyl, diisodecyl, and butyl, benzyl; depending which ester group is cleaved the butylbenzylphthalate gives rise to two monoesters, butyl or benzyl. Further metabolism of the monoesters occurs via omega and omega-1 oxidation of the remaining aliphatic side chain. Measurement of these metabolites is planned in our laboratory. Schmid and Schlatter (11) found that DEHP taken orally by two volunteers (30 mg each) was excreted in the urine at 11 and 15% of the dose. From these and other data they estimated a half-life of 12 hours and concluded that accumulation of DEHP in the body is not likely to occur even though the log partitioning coefficient (Kow) is 4.88. Our analytical approach for measuring the monoesters of the phthalates consists of enzymatic hydrolysis of the conjugate (glucuronide or sulfate), solid phase extraction, and high performance liquid chromatography/mass spectrometry/mass spectrometry utilizing atmospheric pressure chemical ionization in the negative ion mode with quantification by isotope dilution technique using a 4 carbon-13 isotope of each monoester. One difficulty in these measurements is that diisononyl and diisodecyl phthalates are mixtures of several esters due to the impure nature of the alcohols used in their formation. The advantages of our technique are that it is very selective, accurate, precise, and has excellent detection limits. The primary disadvantage is the cost of the equipment and the labeled standards.

We have applied this method to 20 anonymous urine donors from our laboratory. We found detectable levels (greater than 1-5 parts-per-billion) of monoethyl phthalate, monobutyl phthalate, monobenzylphthalate, and mono-2-ethylhexyl phthalate. Most individuals had levels of these phthalate metabolites from 10-100 parts-per-billion. Elevated levels of monoethyl phthalate (about 1 part-per-million) were found in two individuals. These results underscore the need for a thorough evaluation of the American population for exposure to phthalates.

Phytoestrogens

Phytoestrogens are naturally-occurring plant chemicals that exhibit estrogenic activity. They have been deemed to be protective against heart disease, osteoporosis, and certain types of estrogen-mediated cancers. However, concerns have been raised about potential adverse effects in subpopulations (for example, infants on soy based formula) who consume large quantities of phytoestrogens. Phytoestrogens are classified into the following groups:

- Isoflavanoids: found in soy, legumes
- Lignans: found in grains and many fruits and vegetables
- Coumestans: found in clover and alfalfa

Examples of isoflavanoids are daidzein and genistein, which are metabolized to equol and O-desmethylangolensin (O-DMA), respectively; examples of lignans are matairesinol, enterodiol, and enterolactone; coumesterol is an example of a coumestan. The half-life in humans is relatively short for these phytoestrogens; for example, the half-life of equol is reportedly 14 hours; genistein 7 hours; and daidzein 6 hours. The phytoestrogens are primarily excreted in urine as the glucuronide and sulfate conjugates. Our urinary phytoestrogen method for measuring urinary levels of the eight phytoestrogens listed above consists of enzymatically hydrolyzing these conjugates, followed by solid phase extraction techniques, and analysis by high performance liquid chromatography interfaced to atmospheric pressure chemical ionization mass spectrometry/mass spectrometry. Various standards are used to measure the efficiency of the deconjugation (enzyme) hydrolysis step; the efficiency of the remainder of the analytical method is assessed using labeled (deuterated) internal standards. The recovery of the various analytes is excellent, and the method has sufficient lower detection limits to measure phytoestrogen levels in the general population. The highlights of this method are that we get the needed specificity for all eight phytoestrogens. We do this by acquiring chromatographically baseline separation in 10 minutes of all 8 phytoestrogens and by tandem mass spectrometry. As mentioned previously, the method also has the needed sensitivity for measuring general population levels in 2 mL of urine.

We have measured urinary levels of selected phytoestrogens in 200 urine specimens from adults participating in the NHANES III. These 200 participants were not representative of the U. S. population, but they did reside in all 4 regions and in both rural and urban settings. In addition, they consisted of both genders and different races. Dietary information and demographic data are available on these volunteers, but these data are still being assessed. Creatinine values are also available, which allow for normalization based on the urinary creatinine levels. A summary of the results is given in Table 2. The major isoflavanoids, daidzein and genistein, are detectable in the majority of the samples as are the major lignans, enterolactone and enterodiol.

Nonpersistent Pesticides

As part of our Priority Toxicant Reference Range Study of NHANES III we developed a method (12) for measuring 12 analytes, which are pesticides or their metabolites, in human urine. Four of these pesticides are listed in Table 1 as potential endocrine disruptors. These are 2,4-dichlorophenoxyacetic acid (2,4-D), pentachlorophenol (PCP), carbaryl, and parathion. We assessed exposure to these pesticides in a subset of about 1000 adults who participated in NHANES III. Urinary levels of 2,4-D are indicative of recent exposure to 2,4-D or one of its ester derivatives. Likewise, urinary levels of pentachlorophenol indicate direct exposure to that compound although it is believed that small amounts of pentachlorophenol may result from exposure to lindane. However, that is not the case for the nonspecific metabolites, 4-nitrophenol, 1-naphthol, and 2-naphthol. 4-Nitrophenol can result from exposure to a variety of environmental chemicals including ethyl parathion, methyl parathion, EPN, or nitrobenzene. 1-Naphthol is a metabolite of carbaryl and naphthalene whereas 2-naphthol is a metabolite of only the latter compound. Urinary levels from the NHANES III subpopulation are given in Table 3. Exposure data on two other nonpersistent pesticides that give rise to specific metabolites are also listed in Table 3. These are atrazine mercapturate, from atrazine, and 2,4,5-trichlorophenoxyacetic acid (2,4,5-T). Atrazine mercapturate was measured in a subpopulation in NHEXAS and 2,4,5-T in children of Jacksonville, Arkansas (14). These data showed that most of the sampled individuals had nondetectable levels of 2,4-D; 2,4,5-T; and atrazine, but 4-nitrophenol

Table II. Urinary levels[a] of selected phytoestrogens in subset of adults from NHANES III

Phytoestrogen	N	# of N.D.	Range (ppb)	Mean (ppb)	Median (ppb)	LOD[b]
Daidzein	199	30	N.D.-15,900	317	73.8	9.3
Genistein	199	20	N.D.-4,140	35.8	8.9	1.1
Equol	199	45	N.D.-1,840	49.3	4.6	0.3
O-DMA	199	40	N.D.-7,290	129	34.8	4.2
Enterolactone	197	10	N.D.-5,580	512	209	3.3
Enterodiol	198	7	N.D.-2,510	62.8	27.0	0.5
Matairesinol	199	5	N.D.-91.8	14.5	8.6	0.2
Coumestrol	199	190	N.D.-13.1	1.0	N.D.	1.5

[a]Non-detectable (N.D.) values calculated as one half the detection limit.
[b]LOD = Limit of detection

Table III. Human Urinary Levels of Potential EDCs: Nonpersistent Pesticides

Pesticide	Analyte	N[a]	% Positive	Range (ppb)	Median (ppb)	Population
2,4-D	2,4-D	983	1	<1-37	<1	NHANES III
Pentachloro-phenol	PCP	951	64	<1-55	1.5	NHANES III
Carbaryl	1-naphthol	983	86	<1-2500	4.4	NHANES III
	2-naphthol	977	81	1-88	3.4	NHANES III
Parathion	4-nitrophenol	974	41	<1-63	<1	NHANES III
Atrazine	Atrazine Mercapturate	529	0.19	ND-1.5	0.5 (det limit)	NHEXAS
2,4,5-T	2,4,5-T	197	0.51	N/A	N/A	Jacksonville

[a]N = number of human urinary samples analyzed

was found in about one half of the urine samples analyzed. Pentachlorophenol levels were measured in over one half of the specimens, but as noted by Hill et al (13), exposure to PCP is probably decreasing. Because both 1-naphthol and 2-naphthol were detected in a high percentage of the samples, naphthalene, not carbaryl, is probably the origin of these two metabolites.

Summary

Although most research efforts regarding environmental endocrine disruptors have been directed to the chemicals with long half-lives in animals and in the environment, attention should also be focused on those that do not exist as long in these matrices. These chemicals could also be considered "persistent" in these matrices if they are continually released into that matrix. Therefore, assessments on environmental endocrine disruptors should include the toxicity data and exposure estimates for these nonpersistent chemicals as well as for the persistent environmental chemicals. Reference ranges in human specimens for these nonpersistent chemicals are therefore useful in assessing human exposure to these potential endocrine disruptors.

Literature Cited

1. Needham, L.L.; Pirkle, J.L.; Burse, V.W.; Patterson, Jr., D.G.; Holler, J.S. *J. Exposure Analysis and Environ. Epidemiol., Supplement 1* **1992**, 209-221.
2. Colburn, T.; Von Saal, F.S.; Soto, A.M. *Environ. Health Perspect.*, **1993**, *101*, 378-384.
3. Needham, L.L.; Ashley, D.L.; Patterson, Jr., D.G.; Pirkle, J.L; Sampson, E.J. *Toxicol. Letters* **1995**, *82/83*, 373-378.
4. Fingerhut, M.A.; Halperin, W.E.; Marlow, D.A.; Piacitelli, L.A.; Honchar, P.A; Sweeney, M.H.; Greife, A.L.; Dill, P.A; Steenland, K.; Suruda, A.J. *N. Engl. J. Med.* **1991**, *324*, 212-218.
5. Needham, L.L.; Patterson, Jr., D.G.; Burse, V.W.; Paschal, D.C.; Turner, W.E.; Hill, Jr., R.H. *Toxicol. and Industrial Health* **1996**, *12*, 507-513.
6. Lucier, G. W.; Schecter, A. *Environ. Health Perspect.* **1998**, *106*, 623-627.
7. Lucier, G. W.; Needham, L.L. *Environ. Health Letter* **1998**, *37*, 127-128.
8. Sexton, K.; Kleffman, D.E.; Callahan, M.A. *J. Exposure Analysis and Environm. Epidemiol.* **1995**, *5*, 229-232.
9. Pirkle, J. L; Brody, D.J.; Gunter, E.W.; Kramer, R.A.; Paschal, D.C.; Flegal, K.M.; Matte, T.D. *JAMA* **1994**, *272*, 284-291.
10. Wine, R.N.; Li, L.H.; Hommel Barnes, H. L.; Gulati, D.K.; Chapin, R.E. *Environ. Health Perspect.* **1997**, *105*, 102-107.
11. Schmid, P.; Schlatter, C. *Xenobiotica* **1985**, *15*, 251-256.

12. Hill, Jr., R.H.; Shealy, D.B.; Head, S.L.; Williams, C.C.; Bailey, S.L.; Gregg, M.; Baker, S.E.; Needham, L.L. *J. Anal. Toxicol.* **1995**, *19*, 323-329.
13. Hill, Jr., R.H.; Head, S. L.; Baker, S.; Gregg, M.; Shealy, D.B.; Bailey, S.L.; Williams, C.C.; Sampson, E.J.; Needham, L.L. *Environ. Res.* **1995**, *71*, 99-108.
14. Hill, Jr., R.H.; To, T.; Holler, J.S.; Fast, D.M.; Smith, S.J.; Needham, L.L.; Binder, S. *Arch. Environ. Contam. Toxicol.* **1989**, *18*, 469-474.

Author Index

Subject Index

C

D

E

G

H

I

L

R

S

T

Printed in the United States
2291